# Estuarine Management and Quality Assessment

# Estuarine Management and Quality Assessment

Edited by

**J. G. Wilson**
Trinity College
University of Dublin
Dublin, Ireland

and

**W. Halcrow**
Welsh Water Authority
Bridgend, Wales

PLENUM PRESS • NEW YORK AND LONDON

Library of Congress Cataloging in Publication Data

Main entry under title:

Estuarine management and quality assessment.

Includes bibliographies and index.

1. Estuarine area conservation–Congresses. 2. Estuarine pollution–Environmental aspects–Congresses. 3. Estuarine pollution–Congresses. 4. Environmental impact analysis–Congresses. I. Wilson, J. G. (James Gow) II. Halcrow, W.

QH75.A1E88 1985 333.91′64 84-28935

ISBN 0-306-41911-4

Proceedings of an interdisciplinary symposium on Estuarine Management and Quality Assessment, held September 20–22, 1982, in Dublin, Ireland


A Division of Plenum Publishing Corporation
233 Spring Street, New York, N.Y. 10013



Printed in the United States of America

## PREFACE

This volume presents the papers given at the meeting "Estuarine Management and Quality Assessment" held at Trinity College, Dublin in September 1983. The meeting was organised by the Environmental Sciences Unit, TCD and the Estuarine and Brackish-Water Sciences Association and was attended by delegates from the United Kingdom, Ireland, France, Holland and Portugal.

The theme for the meeting was deliberately chosen in an attempt to bring together the various disciplines and interests in the field of estuarine quality, and the extent to which this objective was met may be judged from the diversity of the contributions. The meeting was arranged under three sub-headings which could be roughly summarised thus:-

a) Modelling - as representative of the engineering hydrodynamicist approach.

b) Impact - as representative of the biologist-type approach.

c) Management - attempting to reconcile the different approaches to, and demands on, the estuary.

In fact, since most contributions overlapped into the other two categories, the order in this volume has been changed somewhat from the meeting, and we feel that this book may be fairly presented as one unit. Contributions for which it was not possible to give full transcripts have been included, for completeness, as abstracts. In each session, an invited speaker summarised the state of the art, and the other speakers followed up by describing some new approach to or refinement of the problem. The speakers discussed specific questions about their presentations, and a final closing discussion was held, out of which arose the following points:

a) The meeting stressed the value of interdisciplinary meetings of this kind, particularly as it gave the different disciplines an insight into the problems facing the others and the way in which the latter would like them to be approached.

b) There was considerable interest in the integration of different models, especially hydrodynamic and ecosystem models, as a tool for estuarine management. The importance of being able to quantify effect or changes with pollutant load alterations was stressed.

c) The uses of indices firstly as a communication tool for administrators and public alike and secondly as a focus for scientific investigation received widespread support.

The final point, endorsed by all, was that this meeting was by no means the final word, but should rather be regarded as a stimulus to further work. More information should lead to better ideas and to improved (and easier) management and would be of benefit all round to all users of estuaries, whatever their particular interest.

In conclusion, the organisers would like to thank Trinity College and Bord Failte for their hospitality and the contributors and, most importantly, the Chairman for their part in the discussions. The care and courtesy of Mr. J. Matzka and Dr. R.H. Andrews of Plenum Press and Ms. B. O'Neill and Mr. P. Craig of the Maths Dept., T.C.D. in the production of this volume were greatly appreciated. This book was produced with the NROFF document processing package on the UNIX* operating system.

*UNIX is a trade mark of Bell Laboratories

CONTENTS

# ESTUARINE MANAGEMENT
# AND
# QUALITY ASSESSMENT

# INTRODUCTION

Historically estuaries have played an important part in the development of towns and cities. The use of ships for trade and transport meant that settlements naturally grew up around those estuaries offering both safe havens and easy access to the agricultural or industrial hinterland. The growth of major industries dependent on sea transport for bulk supply of raw material and distribution of the finished product has resulted in heavy concentrations of such industry in the upper reaches of many estuaries. The most significant of these industries are petroleum refining, petrochemicals, metal refining and food and drink production. All of these industries produce large quantities of potentially polluting effluent. Coupled with the sewage arising from the expanding populations these effluents caused a massive deterioration in the quality of many estuaries during the latter half of the 19th and the first half of the 20th century.

Pollution was so severe in some cases that the resulting stench and threat of typhoid and cholera epidemics forced municipal authorities to take action. Both London and Glasgow constructed primary sewage treatment works with disposal of the sludge to sea but nothing was done in cities such as Liverpool and Newcastle. By the 1950's most estuaries in the proximity of ports and cities could no longer support fish and many were virtually open sewers. Even the benefits of earlier improvements on the Thames and Clyde had been lost, overwhelmed by population growth.

The improvement in social conditions since 1950 has been accompanied by a growing concern for the environment. In the United Kingdom and other countries this has been evidenced by a series of statutes for the prevention of pollution and the setting up of the necessary mechanisms for the enforcement of this legislation. Major improvements have occurred as a result and, taken overall, deterioration has halted. Gross organic pollution is much diminished and fish are returning to many industrialised estuaries. The point has now been reached where, within the foreseeable future, virtually all estuaries will have achieved the minimal desired standards of quality, that is they will support the passage of migratory fish and they will not cause visual or olfactory offence. The design and extend of the remedial measures has in some cases been based upon careful scientific observation and prediction of the results of changes in inputs but generally speaking the work has been empirically based and depends largely upon straightforward application of conventional sewage treatment or relocation of discharges to achieve higher environmental standards.

Now that basic standards have been, or soon will be, achieved the most immediate problem is to decide what the ultimate standards of water quality should be. Should estuaries be returned to their original pre-industrial condition or should they be maintained at a level just sufficient to support migratory fish and to allow reasonable amenity use? The answer is probably somewhere between these extremes; estuaries should be reasonably "natural" but economic considerations demand that they also serve as assimilative and transport mechanisms for effluents.

In order that an estuary can both sustain a fishery and also serve as a disposal route for effluents requires a knowledge of the pollutant concentrations which the desired fishery can tolerate or, in the case of oxygen, the minimum concentration necessary to sustain the desired activity in fish; these are normally termed Environmental Quality Standards (EQS). A considerable amount of work has been done on EQS for the most damaging effects associated with domestic sewage discharges, principally deoxygenation and toxic concentrations of unionised ammonia and it is now largely a matter of judgement in the application of standards appropriate to the needs of particular estuaries.

In recent years attention has focussed upon more persistent pollutants, such as organohalogens, and certain toxic trace metals. The response of the flora and fauna to these contaminants is not well known and there is considerable difficulty in selecting appropriate standards. Attempts have been made by both international and national agencies, notably the EPA in the United States and the EEC Commission, but the standards are somewhat arbitrary. Nevertheless guidance does exist which allows some measure of judgement to be applied to the control of discharges of these substances.

Apart from national and international regulations and guidance on acceptable levels of contaminants in the estuarine environment it is also necessary that the overall impact of pollution be assessed in individual estuaries and, wherever possible, it is desirable that significant contaminants be identified so that remedial action can be taken. This can be approached very simply by chemical analysis or through biological assessment. A number of recent studies are described in this volume and it is apparent that a variety of well proven techniques are now available for the overall assessment of the impact of pollution.

The application of environmental quality standards in the control of discharges requires that the existing distribution of the pollutant be known and that the processes controlling that distribution be understood. It is also essential that the quantities discharged be known and that quantified causal relationships can be established between individual discharges and the observed distribution in the estuary. In the complex physical, biological and chemical environment of an estuary this is extremely difficult and could only be achieved at unacceptably high costs. As an alternative relatively simplistic representations of the processes involved can be derived and numerical models applied to synthesise the information available into a reasonable simulation of the prototype. This has been done successfully for a variety of substances and processes in estuaries and modelling is now a well established technique in pollution control. The overall philosophy and current practices of modelling are discussed at length by O'Kane in this volume, while specific examples of recent advances in modelling specific substances and the controlling processes are given by several contributors.

The production of models by estuarine quality often seems to be regarded as an end in itself, offering a complete solution to pollution problems. This is an entirely erroneous view, the true objective must always be the formulation and implementation of a practicable management strategy. Modelling is an essential part of this but there are many other equally important factors in the process. Even a full understanding of all estuarine processes and the ability to predict the consequences of future change is not in itself sufficient to guarantee action nor does it necessarily allow planning of future utilisation of the estuarine system. In order to ensure remedial action it is generally necessary that the costs be justified in conjunction with other priorities, and only the most seriously affected estuaries will merit attention. So that some order of priority can be assigned considerable effort has been put into the development of quality classification systems for estuaries. The exercise carried out in the United Kingdom is described by Wood and Portmann and a more refined system developed for application generally is discussed in detail by Jeffrey et al. Such classificatory systems also serve to demonstrate to Government the effects of their policies and can exert a powerful influence upon future action.

The ability to allocate multiple discharge consent conditions within a finite estuarine capacity to absorb pollutants, while working towards a desired quality objective, must be the desired goal of estuarine management. This would be difficult enough scientifically, but it becomes almost impossible when financial and socio-economic factors are taken into account. The need for investment in industrial expansion and renewal is almost always accorded a much higher priority than the protection of estuarine water quality.

The costs of reducing the contaminants in a discharge effectively to zero are so high as to make tis a unjustifiable policy, even where estuarine quality is already well below the desired standard. This means that some degree of contamination must always be accepted. Where the quality is above the desired standard then obviously capacity exists to absorb the polluting load and it would be foolish not to utilise some part of this capacity.

The art of effective estuarine management is to either maintain or improve the quality of an estuary while accepting discharges at minimum cost to the economy. This requires the "manager" to balance a host of competing interests and satisfy all needs, with the minimum of compromise. Current practice in this art is discussed by Mackay and possible options for formalising future planning are put forward by Nicholas.

There is no unique solution to successful estuarine management. It is essentially a practical science governed to an extent by political objectives which will be applied in a infinite number of ways to achieve better environmental conditions in estuaries. Nevertheless it should not be dismissed as something which can only be achieved as a product of experience, or as a consequence of dealing with events as they arise. The application of scientific method together with a desire to manage and plan for the future should be the conscious aim of all those responsible for the quality of estuaries.

# CONFLICTING VIEWS ON THE ROLE OF MODELS

J. Philip O'Kane

Department of Civil Engineering
University College Dublin
Earlsfort Terrace
Dublin 2

## ABSTRACT

This paper addresses the problem that engineers and scientists view models in different ways. The different objectives of both groups are sufficient to explain this conflict. This thesis is illustrated with appropriate models of estuaries.

## INTRODUCTION

The use and role of mathematical models in studying and managing the ecology of the coastal zone is a subject of controversy. Take for example the following quotations attributed to some well-known biologists.

> 'Many an ecologist believes that a better prediction can be obtained by rapid intellectual appraisal and elaboration of a complex situation than after a tiresome computation based on unreliable measurements and on half-invented constants.' (R. Margaleff, unpublished)

> 'The development of systems modelling will (not) exert as profound an influence on biological thinking as ... statistical analysis' (C.H. Mortimer, 1975)

(Both these quotations are taken from preprinted working papers prepared for the Nato Science Committee 'Modelling of Marine Systems', Offir, Portugal, June 4-8, 1973.)

'Process models are interesting and ... well accepted, ... by contrast, ecosystem models are controversial' (T. Platt et al, 1981. See references).

'The purpose of ecological modelling is to teach engineers ecology' (Anonymous biologist)

This paper attempts to clarify some aspects of this controversy by asking how and why scientists and engineers use models. For a more general review of particular ecological models in estuarine management, the reader is referred to Longhurst (1978), Nihoul (1975), Platt et al (1981) and O'Kane (1980).

The arguments in the paper are presented in the form of a dichotomy between the aims of the decision-maker and the scientist.

## MODELS FOR ENGINEERS AND MANAGERS

A widely held view of decision-making (de Neufville and Stafford, 1971; de Neufville and Marks, 1974) can be summarised in the following procedure

1. Definition of objectives, goals or ends

2. Translation of these ends into measures of effectiveness or performance indices.

3. Generation of alterative courses of action which span the range of choice, open to the decision-maker.

4. Evaluation of each alternative by estimating its performance or effectiveness in achieving each objective.

5. Selection by the decision-maker of one alternative.

6. Implementation of the decision.

This procedure, which originated in Operations Research, Systems Analysis, Management Science and the Theory of Engineering Design is not a description of how people actually make decisions. It is rather, a prescription for advising a model decision-maker, who seeks the best advice on how to answer the fundamental questions, why act at all? Why act now? Why act in this way?

The clarity of the distinction between ends and means is matched by a division of labour and responsibility between the decision-maker or client on the one hand and his advisers on the other, whether they are engineers, scientists, analysts or managers. The decision maker plays a major role in steps 1 and 5 and a lesser role in step 3.

Models are the intellectual tools which are used in step 4 by the engineer or systems analyst in order to evaluate alternatives. The results are sometimes presented as a cost-effectiveness table, score-card, or impact matrix as shown for example, in Figure 1 (a) for a hypothetical sewage outfall. Each entry is made using a model of the consequences of each alternative course of action. The choice of measures of effectiveness and of the alternatives to be explored, depends on the objectives of the decision maker. They merit very careful study before any modelling activity is initiated. If they change, then the structure of the score-card changes and consequently the models which are needed, may change also. Figure 1 (b) shows a different score-card, reflecting an increased concern with the Public Health objective. Clearly different models are required to fill in this table.

It is important for the decision-maker that some indication also be given of the degree of confidence in each entry in the table. This will inform his final selection, step 5, which will depend on his attitude to uncertainty and risk, the consequences of a wrong decision and the relative advantages and disadvantages of each alternative.

Engineers often believe that it is easier to predict the relative differences between alternatives than the absolute outcomes. This is achieved by careful calibration or tuning of the model in order to reproduce the present situation which corresponds to the 'do nothing' option. Changes from this state are then examined. Whereas the scientist may be concerned with the absolute accuracy of a predictive model, the decision-maker may well be satisfied with relative accuracy.

An estuary is a open-access common-property resource. Hence the management objectives in step 1 will arise from the sectional interests of the users of the resource. The clarification of existing or future conflict of use or abuse becomes the major purpose of the score-card and of the models which are used to construct it.

These objectives may also be arranged in a hierarchy under headings such as Public Health, Commerce, Recreation, Ecology, and terminating in particular issues such as pollution of fish quotas. The ecological objective recognises (a) that living resources when conserved, are renewable and (b) the strategic importance of preserving and promoting genetic diversity in order to secure sustainable yields of food, fibres, drugs, feedstock, fuel and research material from the biosphere, both now and in the future.

In this approach to modelling, there is no requirement for a single all-embracing model. On the contrary, each objective or problem area may require its own individual model or models. And these need be no more complicated than the minimum which is required to

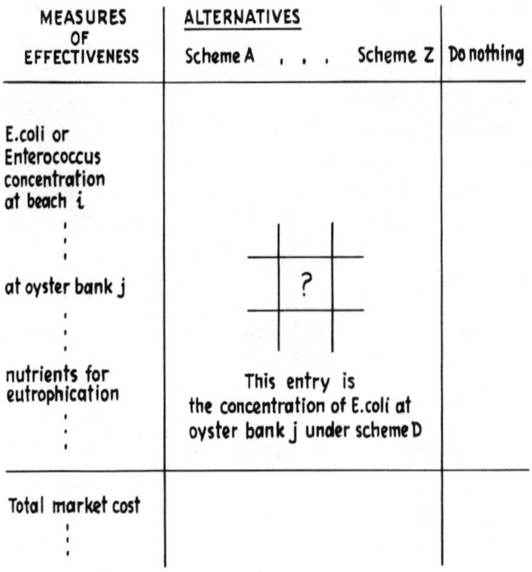


Fig. 1(a)Score card or cost-effectiveness table. (See also text).

fill-in a entry in the score-card with an accuracy acceptable to the decision-maker and his advisers. Models play a different role in the Sciences.

## MODELS FOR SCIENTISTS

The previous section began with a prescription for advising a model decision-maker. There is a similar prescription for a model scientist, namely, the procedure

1. Choose a difficult problem or question as a scientific objective, the answer to which is likely to be of value, as a predictive or manipulative tool for the the advance of science.

2. Answer the question or solve the problem using the scientific method in steps 3 and 4.

3. Generalise the results of previous observations and experiments to form working hypotheses (inductive inference).

4. From these hypotheses deduce the outcome of a further experiment, testing the hypotheses (deductive inference).

5. Publish the results and repeat the procedure.

| MEASURES OF EFFECTIVENESS | ALTERNATIVES Scheme $\alpha$ . . . . . . Scheme $\omega$ | Do nothing |
|---|---|---|
| Visible floating matter at beach i | | |
| ⋮ | | |
| Increase in the rate of incidence of non-specific gastro-enteritis (GI) among bathers at beach j | | |
| ⋮ | Entries to be filled in using appropriate models | |
| Increase in the GI rate of disease among consumers of shell-fish at bank k | | |
| ⋮ | | |
| Impact on fisheries | | |
| ⋮ | | |
| Net number of jobs created | | |
| ⋮ | | |
| Total market cost | | |

Fig. 1(b)An alternative score card. (See also text).

The repeated application of steps 3 and 4 will disprove some hypotheses and accumulate a weight of evidence in favour of others. The final result is often an elaborate theory, supported by empirical data, which permits the prediction of future events in the system of interest.

A system is a set of interacting parts. System function is synonymous with these interactions. In the study of ecosystems, therefore, the first problem is the choice of entities or parts of the system and the second problem is the description, understanding and prediction of their interactions, the 'function' of the ecosystem. Models are used in the elaboration of both these problems.

There are many different ways of classifying the parts of an ecosystem, as living or non-living, as a taxonomic hierarchy (family, genus, species, population, individual), as life-forms or structures (trees, shrubs, herbs, bryoids), as functional entities in a trophic pyramid (herbivores, carnivores, omnivores decomposers) as gene pools and so on. The only constraint on the method of classification is (a) that it be unambiguously defined and (b) that it facilitate the measurement of those interactions which are of interest. The initial question (step 1) provides the criterion for deciding on the most appropriate classification. Different methods of classification lead to different families of models which evolve with the progress of research.

Figure 2 provides an example which illustrates this procedure (Nixon, 1981). The problem of nutrient cycling in coastal marine ecosystems has been studied by marine scientists for a long time. The major questions have been and still are. Can they be quantified and predicted? How do they interact? What are the effects of nutrient enrichment by sewage effluent? How important are salt marshes in these cycles? The sets of ecosystem-parts and their interactions, which are relevant to such questions, are often called _conceptual models_. As we shall see, they always form the basis for more elaborate models of the system's dynamics.

Figure 2 contains two sequences of conceptual models, one for macrophyte-based systems, the other for plankton-based systems, both originating in the "great cycle" model prior to 1950. H.T. Odum's (1971) picture language is used to classify the parts of the ecosystem as standing stocks of primary producers (bullet- nosed boxes), self-maintaining communities (hexagons) and other forms of matter (hemispheres with conical roofs). The lines show the interactions between the parts as flows of nutrients. The hatching indicates flows of lesser significance. The models also imply that stocks at any point in time result from the accumulation of only those net flow rates which are shown.

A complete, empirical verification of the currently accepted conceptual model would require the measurement by sampling of all stocks _and_ _all_ _flows_ during a fixed interval of time. The null hypothesis - that the dynamic nutrient budget is in balance at all times except for sampling error - could be tested, provided the standard errors were also estimated. This would require very considerable resources and has never been attempted. The changes which have occurred in successive conceptual models have been brought aout by laboratory and field experiments confined to small parts of different systems in different parts of the world. Radioactive and stable isotope tracer experiments have been very powerful techniques in such studies. Nevertheless, contradictory findings can be found in the literature. Nixon (1981) counsels that scientists 'should be modest in assessing the present state of certainty with regard to many of the questions now under study'.

## MODELS WITH FEEDBACK PROCESSES

A complete verification of a conceptual model by measurement alone, the purely empirical position, is not the only possibility. A great deal of laboratory research has also been carried out, attempting to correlate each flow rate with the stock levels (or state variables) which influence it. These experimentally derived relationships are called process models in a recent survey by Platt et al, (1981). Two remarks can be made.

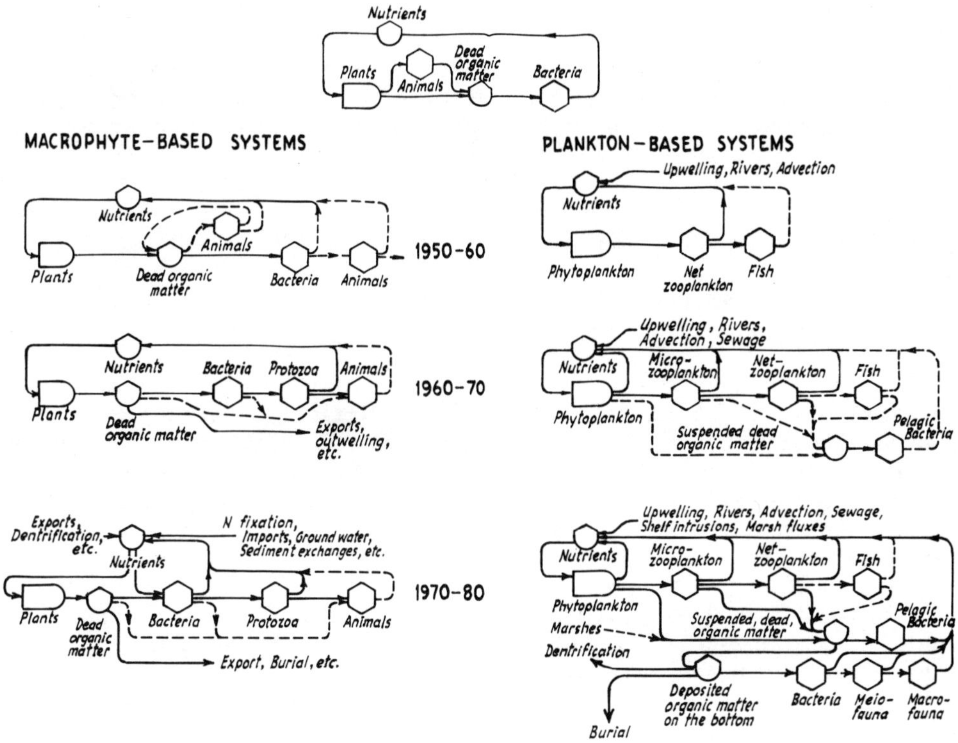


Fig. 2 Changing views of remineralization and nutrient cycling in coastal marine systems. The solid lines represent flows that were thought to be important or seemed to receive a lot of attention. (Taken from Nixon, 1981). (See also text).

In the first instance, they represent an expansion of the conceptual model because a further set of interactions between the parts is hypothesised. Odum's (1971) picture language contains a 'work gate' symbol with which to represent the influence of those stock levels which control the flows. These are often excluded for reasons of clarity or uncertainty, as in Figure 2. In the second instance, these regressions of rates on levels, allow one to write down and close a set of first-order ordinary differential equations, which are based on the principle of conservation of mass or energy. Closed equations can be marched forward in time, continuously updating the accumulated stock levels in response to current net flow rates and readjusting the flow rates from their causal relationships with the stock levels. The conceptual model now has a closed feed-back structure of alternating flow rates and stock levels. Where there is positive feedback there is growth; where there is stability, there is negative feedback (See also O'Kane and Dowley (1978)).

The solution of these equations yields the variation in time of all stock and rate variables. The only information which is required to do this, is (a) initial values of all stock levels (b) the parameter values in all the regression equations of rates on levels, and (c) flow rates across the boundary of the system, e.g. rivers or solar radiation, which force the system.

At this juncture the model is usually called a complex simulation model. See for example Kremer and Nixon (1978). Such models should be applied to replicated field or laboratory micro- or mesocosms, before they are combined with hydrodynamic models and applied in the field.

From this discussion a number of conclusions may be drawn. The original scientific question or objective is all-important. It organizes the evolution of the conceptual model which guides the work of the community of scientists who are committed to it. As long as scientists believe that any particular conceptual model is incomplete, they are likely to view with scepticism, any management models derived from it.

However, most managers are accustomed to making decisions on the basis of incomplete or uncertain information. The consequent disagreement between scientist as scientist and engineer as manager derives from their different objectives. A clarification of objectives will clarify the disagreement, but may not resolve it. The scientist uses models as instruments of management and action.

## THE DEFINITION OF A MODEL AND ITS VERIFICATION

A model of a prototype system or object is a simplified set of interacting parts, and associated hypotheses which can be manipulated, in order to answer a question, solve a problem, or make a prediction about the object.

When the model assumptions can be stated as equations, we can manipulate them using standard mathematical techniques in order to find the Since computers are machines for manipulating symbols, we often make use of them and speak of mathematical computer models.

The interactions are often distinguished as interrelated inputs and outputs to and from the parts of the model object and its environment. The input/output relationships require the specification of 'causal' and 'residual' hypotheses which may form the basis of a statistical treatment of measured data.

When controlled experiments on the prototype object are possible, the use of the randomisation procedure isolates the disturbing effect of uncontrolled factors in the residual errors and allows the application of exact and efficient techniques of parameter estimation and hypothesis testing. This is often impossible or very difficult in the environmental sciences. See for example the plea of Hall (1971) for more controlled field experiments deliberately stressing the ecosystem. At present, passive observation is the rule. Hypothesis testing in these circumstances is inexact and difficult. Wold (1956) lists three methods of testing hypotheses where non-experimental or observational data have been collected: (1) using additional data from other studies, (2) using a priori arguments, and (3) using predictive tests with new measurements. There is sufficient latitude in this to ensure a lack of consensus. The major difficulty is that different causal models can often be made fit the data equally well.

Mortimer (1975) summarised the role of mathematical models as a means of (a) uncovering and displaying patterns inherent in the data, (b) exposing data gaps, (c) optimising data collection, (d) designing verification experiments, (e) sensitivity testing beyond the measured variability of the data, and concluded that they are no substitute for insights or ideas. But, of course models may stimulate insights through the efficiency with which they encapsulate ideas. See for example, the work of May (1974). They are therefore well suited as a means of communication, step 5 of the procedure for a model scientist.

Models are metaphors. We manipulate them using symbols, words, equations or similar objects to a smaller scale. We seek similarity but not identity between object and model, which is sufficient to fulfil the purpose of the model. Hence models are also instruments of economy in making predictions, solving problems and communicating the answers to questions about a object of interest.

Finally, microcosms and mesocosms are also models since they are 'similar objects to a smaller scale' which can be manipulated physically, chemically and biologically. There are many questions which can only be studied in this way at the present time. They complement mathematical models. Both have their place.

## THE ENGINEER'S VIEW OF MODEL VALIDITY

We shall begin this discussion with a simple example. Suppose two steady streams of water flow together to issue from their junction as a third stream. Each contains a common solute. This is the object of interest. We seek the relationships between flow rates and concentrations in the three streams. We present the model first, its symbols and assumptions, then the questions to be answered.

Let $Q_1, Q_2$ and $q$ be the water flow rates in the three streams measured in $(m^3 s^{-1})$ shown in Figure 3. Let $C_1$ and $C_2$ be the concen-

A simplified set of parts

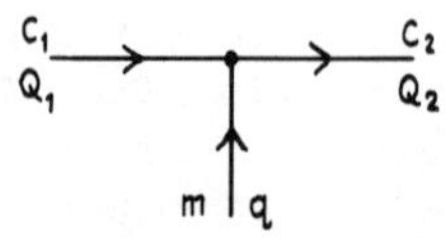


Rate balances between the parts

1. Water $Q_1 + q = Q_2$ $[m^3 s^{-1}]$
2. Solute $Q_1C_1 + m = Q_2C_2$ $[g\ s^{-1}]$

| PROBLEM | GIVEN | FIND |
|---|---|---|
| How does an effluent change the downstream concentration? | $(m, q, C_1, Q_1, -)$ | $C_2 = C_1 + \frac{m - qC_1}{Q_1 + q}$ |
| What effluent will match a water quality standard downstream? | $(-, q, C_1, Q_1, C_2)$ | $m = (Q_1 + q)C_2 - Q_1C_1$ |
| What is the flow in the stream? | $(m, q, C_1, -, C_2)$ | $Q_1 = \frac{(m - qC_2)}{C_2 - C_1}$ |

Fig. 3 A model for the confluence of two streams

trations of solute in the main stream measured as $(g\ m^{-3})$.m is the mass flow rate of solute in the tributary measured in $(g\ s^{-1})$.

By making the assumption that mixing is a conservative physical process, which neither creates nor destroys water and solute, we may write down two flow-rate balances at the confluence of the streams. Both balance equations are given in Figure 3. They contain 6 symbols which can be manipulated in different ways to answer different questions. Let us look at two different sets of questions.

Suppose a Planning Officer wishes to know the change in concentration of a conservative solute in a river due to a proposed new discharge. In terms of our model, the question is 'Given $C_1$, $Q_1$, m and q, what is $C_2$'? A simple manipulation yields the answer in Figure 4. If $C_2$ exceeds the receiving water standard C he may pose the additional question "given C, Q, g and $C_2$, what is m ?". The answer is shown in Fig. 3. By substituting C for $C_2$ in the right hand side, he can find the effluent standard which he must set, in order to meet the receiving water standard.

As a second example, consider a Hydrometric Engineer who wishes to measure the flow in a river by dilution gauging i.e. by injecting dye at a steady rate m and measuring the dilution of the dye downstream. His question is A simple manipulation yields the answer in Figure 3.

Now many scientists would not call this a model because of its simplicity. However it satisfies our definition and its value as an example, will become apparent.

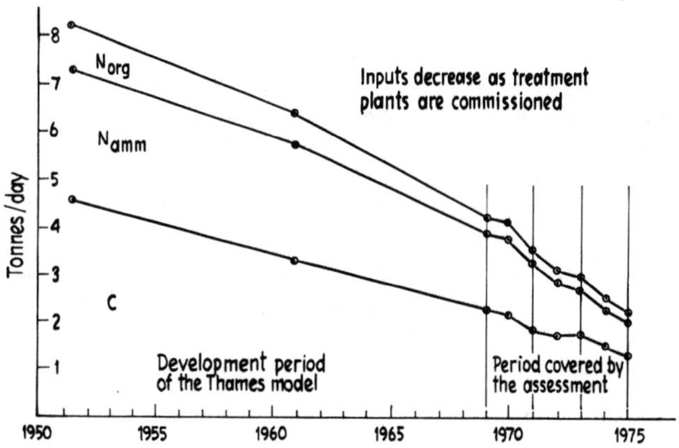


Fig. 4 Effective oxygen demand input to the Thames, showing total input (top line) and contributions of organic carbon (C), ammonia ($N_{amm}$) and organic N ($N_{org}$). From Table 1, Barrett et al (1972). See also O'Kane (1980) chapter 2.

All mathematical models have the form stands for the assumptions of the model which when manipulated, yield the answer or conclusions B. The same model, as we have seen, can yield answers to different questions posed by different people for different purposes. There is therefore, no single position from which to judge the validity of a model.

The Planning Officer, in examining the validity of his conclusions, will be aware of the fact that river flows and effluents are random variables which vary from day to day throughout the year. Steady conditions are not the rule. To overcome this he may use a range of extreme values of Q, C and m in order to inform his judgement. If he is still unhappy about the consequence of a wrong decision, he may extend the model to include the daily variation in Q, C and m. Using the same rate balances as in Figure 3 and an appropriate computer, he can march the state of the river forward in time day by day collecting the statistics of interest to him. Of course he now needs daily measurements of Q,C and m to drive the model.

If the discharge is made to an estuary, the model can be extended to include the variation in concentration in one or more directions and also in various biotic and abiotic compartments. The evolution of the state of the ecosystems to those in Figure 3.

The Hydrometric Engineer on the other hand has a different set of concerns. If changes in flow Q are sufficiently small, during the hour in which he completes his measurements, his major concern will be the control of systematic and random error in the derived flow Q. Random errors can be controlled by repetition and randomisation in the field. Systematic errors can be controlled partially, by repeated comparison against a more accurate method of flow measurement in the laboratory. By introducing assumptions about the random nature of the errors, it is possible to deduce a standard error of the estimate of Q.

Further questions can be asked of this probabilistic model, 'What measurement procedure and resources minimise the cost of attaining a prescribed standard error in Q? Of course the setting of the standard error in Q and the associated minimum costs can only be made at a higher level. A cost-effectiveness analysis of the Planning Officer example could be used to do this.

Clearly the validity of any model cannot be judged without reference to its purpose namely the questions, problems and decisions which are of concern to the model user. This has the great advantage of defining a modelling procedure namely, problem -> model -> data -> validity. The problem statement provides the criterion for deciding what to exclude and what to include in the model. Field and laboratory-data requirements are specified by the model itself.

Two conclusions emerge from this example. Firstly, a range of more and more elaborate models, which progress in different directions from the common model in Figure 3, is available to each decision-maker. Secondly, when they make a decision, it will be informed for better or worse by one of the models in the progression of possible models. The choice is not between a model or no model at all, but rather between several competing models. Which model is chosen, will depend on the penalties which may be incurred by a wrong decision and on the incompletely perceived incremental costs of making the next model in the progression. If there is no problem to be solved or decision to be made, there is no need for a model. This pragmatic problem-oriented view of models is shared by many engineers.

## TWO SCIENTIFIC MODELS

We shall conclude this essay with a brief comparison of two models of oxygen depletion in two grossly polluted estuaries, namely, the Thames (Thames Report 1964) and the Western Scheldt (Billen and Smitz, 1978). Both estuaries receive discharges of sewage in their upper reaches, from London and Antwerp respectively, which are sufficient to produce large anoxic zones.

### The Thames Model

The Thames model was made for the purpose of management, but without the acute pressure for a quick answer which is characteristic

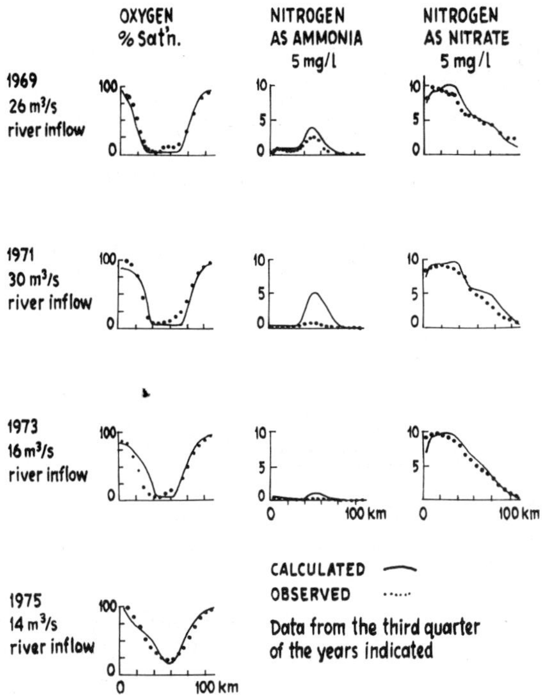


Fig. 5 Assessment of the Thames estuary pollution model. From Fig. 2, Barrett et al (1978).

of engineering decision-making. The scientific objective was to develop a predictive causal model relating discharges of sewage at any point along the Thames estuary to the concentration of oxygen at any other point. A recent review (Barrett and Mollowney, 1978) reports the remarkable success of this model in predicting changes in oxygen, ammonia and nitrate concentrations following the introduction of major treatment works (Figures 4 and 5).

This can be regarded as a large statistically uncontrolled experiment, verifying the causal hypotheses of the model in accordance with Wold's (1956) third method of model verification. Nevertheless, it has been subjected to criticisms, particularly from biologists using a priori arguments, Wold's (1956) second method of model verification. Such critics point out that the Thames model does not agree with ay of the conceptual models in Figure 2. The model does not contain a single stock level or state variable which explicitly represents phytoplankon, zooplankton, bacteria or protozoa. How could the model be 'correct' with these omissions?

It has been found in a very large number of laboratory experiments and tests that the respiration of an adapted microflora feeding on bio-degradable wastes, can be represented in a simple way. The

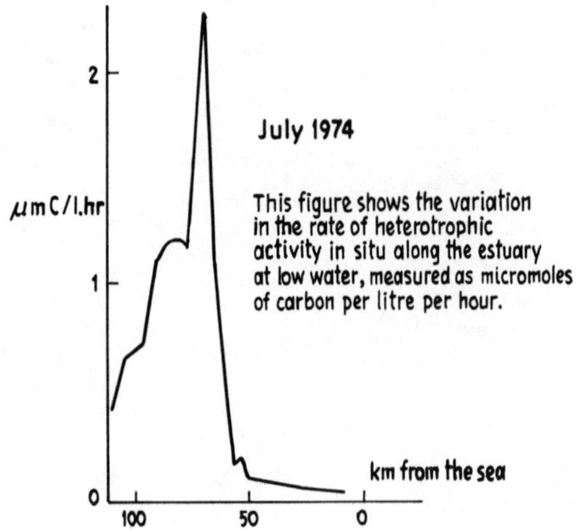


Fig. 6 The forcing function of the Scheldt model. From Billen et al (1976).

rate of respiration depends on the concentration of organic carbon, nitrogen and ammonia in the waste. This rate and its dependence on temperature has been determined for the Thames and for many other polluted bodies of water around the world. For further details see Chapter 2, O'Kane (1980) and the Thames Report (1964). The notable advance made in the Thames Study, was the extension of the findings to an anoxic estuary where oxidants other than oxygen also play a role. This defence of the Thames model is an appeal to Wold's (1956) first method of model verification.

However the disagreement concerning phytoplankton remains. It may be that the net contribution of phytoplankton to the quarterly oxygen budget for the Thames is relatively small in the most polluted zone. For further remarks see Owens (1972) and the ensuing discussion on this point. However, the most interesting criticism has been made by Torpey (1967) based on an examination of historical data from New York Harbour and the Thames Estuary:

> 20 - 150 kg-$O_2$/ha/day, the dissolved oxygen remains substantially constant for the subject 9m deep estuaries, at between 25 and 50 per cent saturation. This zone of the dissolved oxygen plateau is homeostatic in nature, because the symbiotic activities of algae and bacteria are able, with an assist from the atmosphere, to supply enough oxygen to accommodate load variations.' Torpey (1967), page 1808.

Torpey (1967) did not use the available mathematical models in arriving at his conclusion. He might have strenthened his case considerably, if he had. Since the Thames is about to enter Torpey's 'homeostatic oxygen plateau', further reductions in waste loading may

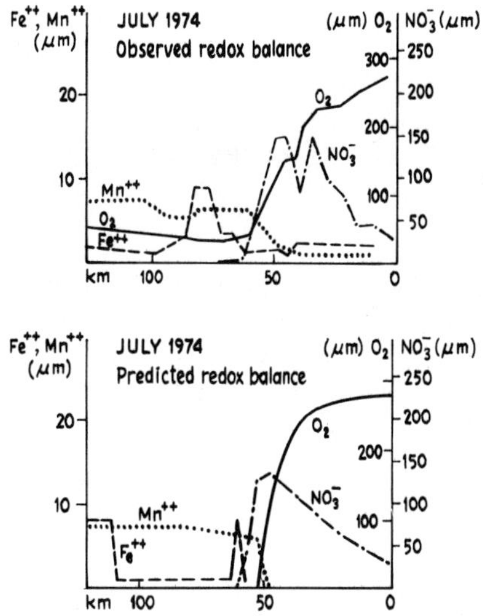


Fig. 7 Comparison of predicted and observe redox balance for July 1974 in the Scheldt. From Billen et al (1976).

only yield a marginal improvement. Before accepting this hypothesis, the Thames model should be used to 'retrodict' the historical data which Torpey used and any mismatch should be examined. If this confirms the hypothesis, then the model should be extended before being used to guide the future management of the estuary.

## The Western Scheldt Model

The model of the Western Scheldt (l'Escaut) was not made for immediate management purposes at all. The objectives were scientific and follow from two observations: (Billen et al. 1976) (a) organic waste is not the limiting factor in heterotrophic activity in anoxic estuaries or muds, and (b) a complete redox balance is necessary to understand the kinetics of their self-purification. Figure 2 does not raise either of these problems. The Thames model recognises them, but treats them parsimoniously.

The scientific objective of the Scheldt model was to understand and predict the redox chemistry of the estuary. The model predicts the distribution of five redox couples (oxygen, nitrogen, manganese, iron and sulphur) along the estuary using the appropriate Nernst equations, thermodynamic data, boundary conditions and perhaps a steady state one-dimensional mixing model (Wollast, 1978).

The redox model is decoupled from the discharges of sewage and its distribution within the estuary by means of an in-situ measurement of the heterotrophic activity along the estuary at low water. The uptake rate of radio-labelled $HCO_3$ ion in the dark, was taken as a semi-quantitative measure of the activity rate. This rate drives the redox chemistry as a electron flux. For this reason, the response of the Scheldt to a decrease in the discharges of waste cannot be assessed with this model. It cannot be used to fill in the entries in a score card, in its present form.

Figure 6 shows a typical measured heterotrophic activity rate which drives the redox chemistry of the anoxic Scheldt. Figure 7 shows a comparison of predictions and observations. It is a unique example of modelling a stressed natural aquatic system using classical chemical thermodynamics.

## CONCLUSIONS

While the estuaries of the Thames and Western Scheldt are similar to each other, different objectives lead to marked differences between their models. This illustrates very well our contention that models depend strongly on their purpose, whether it is to answer questions of interest to science or of interest to management. Throughout the paper we have contrasted these two positions as a dichotomy. The ten years of research in developing the Thames model shows that this is not always the case. Clearly further developments are possible which combine the insights of the Thames and Scheldt models, and Torpey's critique.

This paper has also been written in the hope that some biologists and chemists might be convinced to look more favourably on models as an integral part of research, an approach which is commonplace in the physical sciences. Of course there are bad models, models which have so many parameters that they can be made to reproduce any set of data or which contradict fundamental requirements or which conceal a lack of knowledge. If we are to progress beyond a catalogues of facts and procedures, we must make models.

## REFERENCES

Barrett, M.J. and Mollowney, B.M. 1978. The Thames Model: An Assessment. Progress in Water Technology, No. 10, Nos 5/6, 409 - 416

Billen, G and Smitz, J. 1978. Mathematical model in a highly polluted estuary. In: Hydrodynamics of estuaries and fjords. (Nihoul J.C.J. (ed)) Amsterdam, Elsevier Press, p. 55 -62.

Billen, G., Smitz, J., Somville, M. and Wollast, R. 1976. Degradation de la matier organique et processus d'oxydo-reduction dans l'estuaire de l'Escaut. Chapitre 111 A. In: Project Mer, Rapport Final. Volume 10, l'Estuaire de l'Escaut (Nihoul, J.C.J. and Wollast, R. (Eds)) Brussels, Services du Premier Ministre, Programmation de la Politique Scientifique, p. 101 - 152.

de Neufville, R. and Stafford, J.H. 1971. Systems Analysis for Engineers and Managers. New York, McGraw-Hill.
de Neufville, R. and Marks, D. (Eds) 1974. Systems Planning and Design. Case Studies in Modeling, Optimization, and Evaluation. New Jersey, Prentice-Hall Inc.
Hall, D.J. 1971. The experimental field approach to secondary production. In: A manual on methods for the assessment of secondary productivity in fresh waters. Edmonson, W.T. and Winberg, G.G. (Eds)), London, Blackwell Scientific Publications. p. 210 - 221
Kremer, J.N. and Nixon, S.W. 1978. A coastal marine ecosystem-simulation and analysis. New York, Springer Verlag.
Longhurst, A.R. 1978. Ecological models in estuarine management. Ocean Management, 4, p. 287 - 302.
May, R.M. 1974. Stability and Complexity in Model Ecosystems. 2nd Edition. New Jersey, Princeton University Press.
Mortimer, C.H. 1975. Modelling of lakes as Physico- biochemical systems-present limitations and needs In: Modelling of marine systems, (J.C.J. Nihoul (Ed)), p. 217 - 32. Amsterdam, Elsevier Press.
Nihoul, J.C.J. (Ed) 1975. Modelling of marine systems. Amsterdam, Elsevier Press. Nixon, S.W., 1981, Remineralisation and nutrient cycling in coastal marine ecosystems. In: Estuaries and Nutrients. (B. Neilson and L.E. Cronin (Eds)), U.S.A., Humana Press.
Odum, H.T. 1971. Environment, power and society. New York, John Wiley and Sons.
O'Kane, J.P.J. 1980. Estuarine Water Quality Management with moving element models and optimization techniques London, Pitman Publishing Ltd.
O'Kane, J.P.J. and Dowley, A. 1978. Models for Impact Assessment in Estuaries and the Coastal Zone - An introduction without mathematics. In: Coastal Pollution Assessment. W.K. Downey and G. Ni Uid (Eds)). Dublin, Ireland The National Board for Science and Technology, p. 147-162.
Owens, M. 1973. Sources of oxygen in estuaries. In: Mathematical and Hydraulic Modelling of Estuarine Pollution (A.L.H. Gameson (Ed)), Water Pollution Research Technical Paper No. 13. London, HMSO. p. 25 - 37.
Platt, T., Mann, K.H. and Ulanowicz R.E. (Eds) 1981. Mathematical models in biological oceanography. Paris, Unesco Press.
Thames Report 1964.. Effects of Polluting Discharges on the Thames Estuary, Water Pollution Research Technical Paper No. 11 London, HMSO.
Torpey, W.N. 1967. Response to pollution of New York Harbour and Thames Estuary. Journal of the Water Pollution Control Federation, 39. No. 11, 1797 - 1809. Wold, H.D.A., 1956, Causal Inference from Observational Data: A review of ends and means. Journal of the Royal Statistical Society Series, 119, p. 28 - 61.

Wollast, R. 1978. Modelling of biological and chemical processes in the Scheldt estuary. In: Hydrodynamics of estuaries and fjords. (J.C.J. Nihoul (Ed)) Amsterdam, Elsevier Press. p.63 - 77

# AN INTEGRATED MODELLING APPROACH TO ENVIRONMENTAL QUALITY MANAGEMENT

P.J. Radford

IMER, Prospect Place
The Hoe, Plymouth PL1 3DH
England

## ABSTRACT

The theme of this paper concerns the role of models in integrating research across disciplinary boundaries and their use as an aid to environmental quality management. It adds too the contribution made by Radford et al., in the EBSA Local Meeting on the Tamar Estuary (April 1982) but concerns a different application of the same principles.

Biochemical Oxygen Demand (BOD) models have long been established as a management tool and have been used most effectively in monitoring the recovery of a number of grossly polluted estuaries. Their limitations in predicting oxygen levels in cleaner environments has always been acknowledged, but attempts to include the contribution of primary producers have been limited and largely unsuccessful. Ecological models such as GEMBASE (General Ecosystem Model of the Bristol Channel and Severn Estuary) have tended to be regarded as rather esoteric and certainly too theoretical for practical estuarine management. Recent work at IMER has been concerned with the integration of the best features of both BOD and ecosystem models into a single model which is capable of predicting the transition of a highly polluted estuary into a near normal ecosystem and vice versa. This work, which has been done in collaboration with the Hydraulics Reseach Laboratory has used a three-dimensional hydrodynamic model to compute exchanges between system elements of dissolved and planktonic components of the estuary. A classical BOD model based upon research from the Water Research Centre (Stevenage) has been developed to compute oxygen concentrations but a ecosystem model based on GEMBASE interacts with both of these models to allow for the impact of the living components. Thus two types of model which have always been considered as incompatible have been integrated to produce a practical management tool.

# OBSERVATIONS AND MATHEMATICAL MODEL FOR THE WYRE ESTUARY

A.M. Riddle

ICI Brixham Laboratory
Brixham
Devon
England

## ABSTRACT

The Wyre is a small estuary (22 km in length) in north-west England which flows into Morecambe Bay at Fleetwood. The tidal section is very narrow for about 14 km down from St. Michael's on Wyre and then opens out to become up to 1.3 km wide just above Fleetwood. The estuary is "well mixed" at all states of tide, has a large intertidal volume and virtually empties at low water to leave a meandering channel through the sand banks of the lower estuary. The estuary has been well studied and has been surveyed biennially since 1967.

There are numerous discharges to the estuary, both domestic and industrial, so in order to assess the impact of these effluents a mathematical model has been constructed. The model consists of a hydrodynamic section to compute the current and depth distributions and a mass balance section for predicting salinity or pollutant concentrations. Observations of currents, salinity and dissolved oxygen are presented together with predictions from the model. This model is now being used to assess the relative importance of individual discharges and the effect of changes in position of outfalls.

## INTRODUCTION

The Wyre rises in the Forest of Bowland, flows west then south through Garstang, picking up several tributaries on the way. It continues west through St. Michael's on Wyre, becoming tidal here, and on to Poulton le Fylde. The estuary turns north and widens before reaching Fleetwood and emptying into Morecambe Bay (Figure 1). The

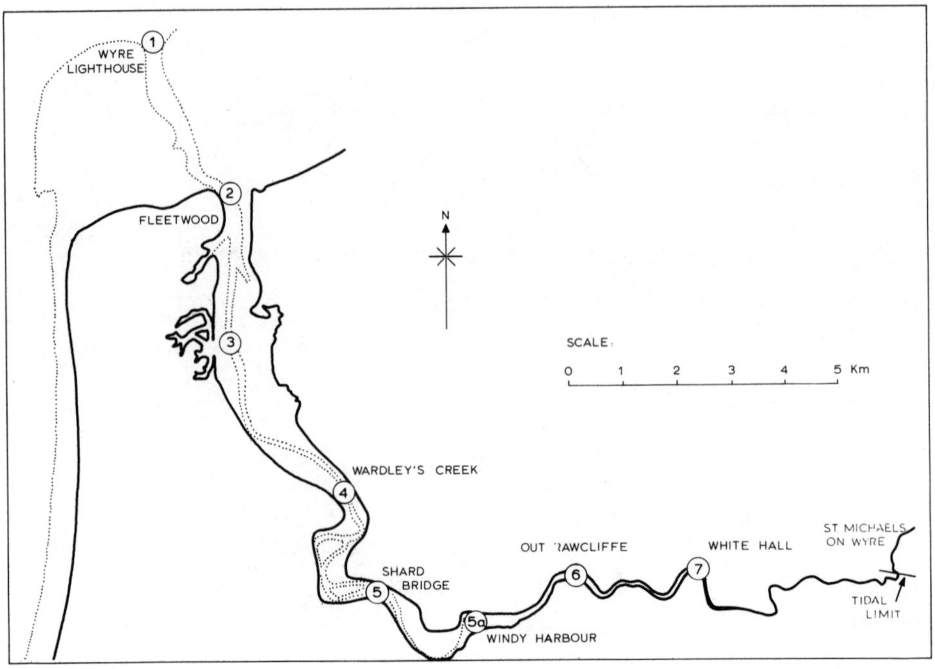


Fig. 1 Wyre estuary. Sampling stations and outfalls.

whole area from Garstang to the sea is very flat (below 100 ft.) with the river meandering across this plain until reaching the lower estuary and the final 7 km to the sea.

The upper estuary, from St. Michael's on Wyre to Shard Bridge (12 km) is above mean tide level and is a river for most of the tidal cycle with a short flood of less than two hours. This part of the estuary has a smooth mud bed and banks. Below Shard Bridge the estuary widens and the mud gradually changes to sand. Inglis and Kestner (1958), described the sediment distribution and changes of position of the low water channel in detail. The estuary bed is aove the low water springs level to within 3 km of Fleetwood, from where it deepens rapidly to 5 m below chart datum in the dredged shipping channel between Fleetwood docks and the Wyre Light. The estuary has been surveyed biennially since 1967 by the North West Water Authority, the Lancashire and Western Sea Fisheries Joint Committee and the ICI Brixham Laboratory.

The surveys consisted of sampling five fixed stations between the Wyre Light and Shard Bridge (Figure 1) until 1977 when two stations further upstream were added to obtain more information on lowered oxygen levels. An eighth station was added in 1979 between the existing 5 and 6 positions. A boat was anchored at each station and measurements taken half-hourly over a full tidal cycle (25 hours

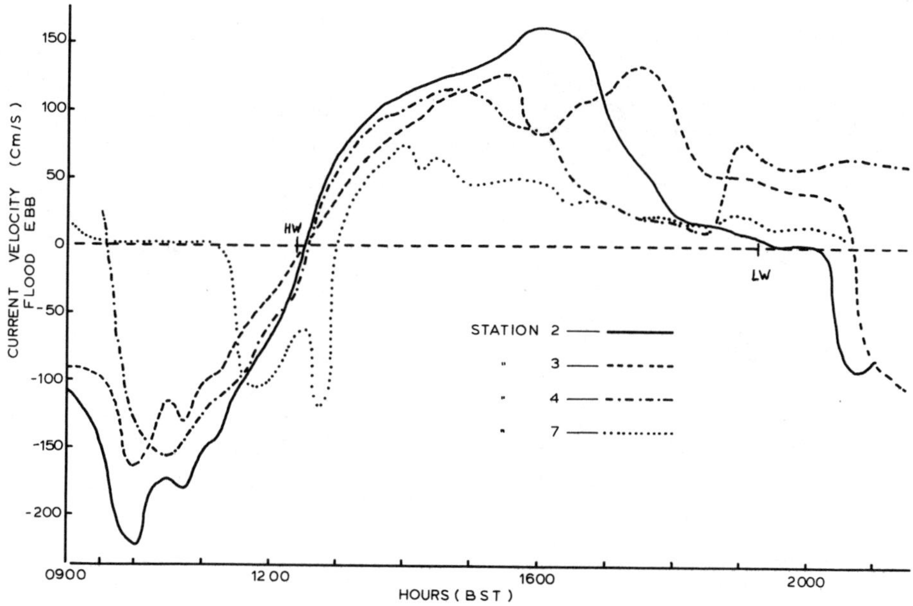


Fig. 2 R. Wyre surface currents.

in 1977). At each sampling time instruments were used to measure current velocity and direction, salinity, temperature, dissolved oxygen and pH at specified depths throughout the water column (surface, 0.5m, 1m, 2m, 4m and then at 3m intervals to the bottom). Additionally water samples were taken for analysis to determine dissolved oxygen, nutrients and heavy metals. The survey periods have covered tidal ranges from neaps to springs and a wide range of fresh-water flows.

## MODELLING THE WYRE

The estuary is well mixed at all states of the tide (maximum recorded salinity differences surface to bottom are 0.1 $^{o}/oo$ ) and salt penetration has been detected within 3 km of the weir at St. Michael's on a spring tide (range 9.3 m), though normally it does not reach station 7. Tidal currents measured in 1979 are shown on Figure 2. These were for a spring tide (range 8.4 m) and clearly show the progression of slack water times up the estuary and the reduced duration of the flood tide at the upper stations. The initiative for modelling the Wyre came after the 1975 survey during which low dissolved oxygen levels were recorded in the middle reaches of the estuary (60%). An explanation was required which led to the development of a model. The model originally considered the estuary from Morecambe Bay to Shard Bridge, but was later extended to the tidal limit at St. Michael's (Figure 1) after cross-section surveys were carried

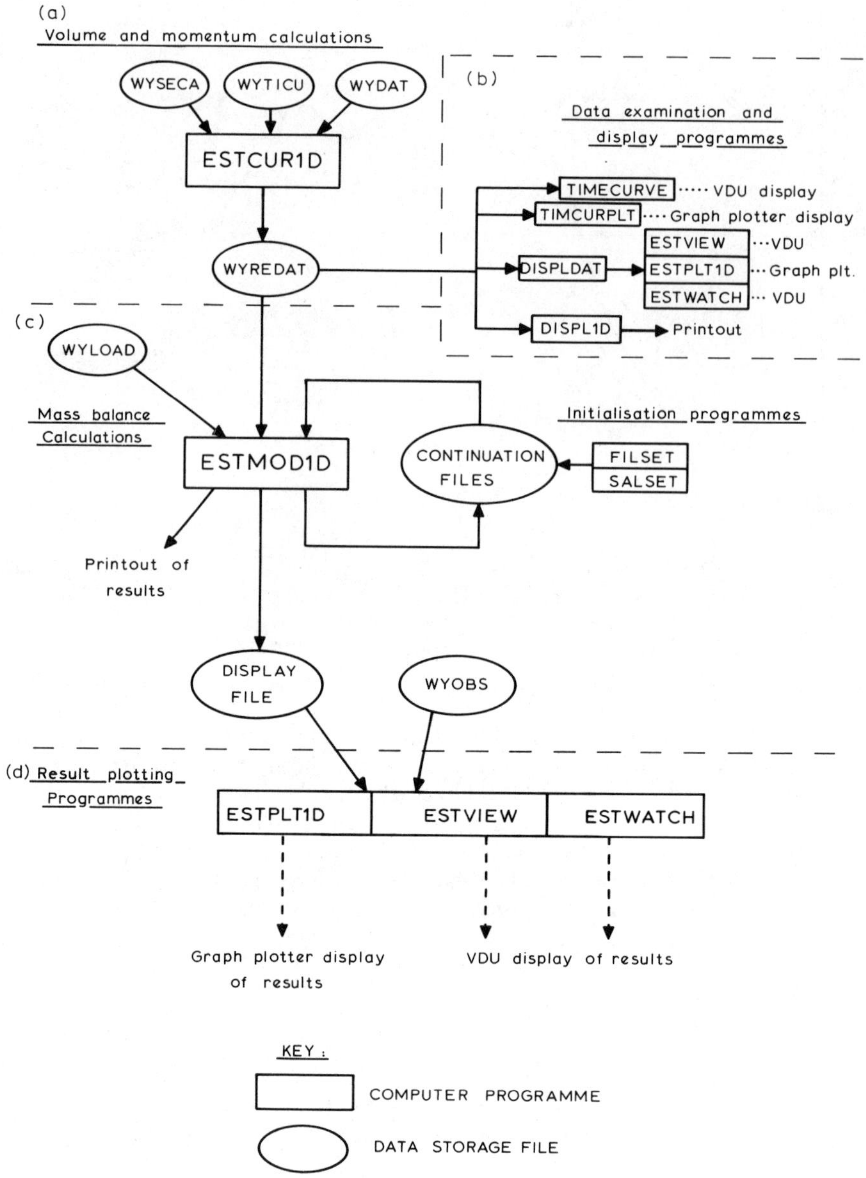


Fig. 3 Modelling system.

out by ICI Brixham personnel in 1977 and 1978. These surveys and British Transport Docks Board soundings covered the whole estuary from Wyre Light to the tidal limit. The model is one-dimensional and time-varying with hydrodymanic and mass balance sections to compute the currents/depths and salinity /pollutant distributions along the length of the estuary. The equations used for the model are the cross-section integrated forms of the Navier-Stokes equations for the continuity (1), momentum balance (2) and mass balance (3) within the estuary.

$$b\frac{\delta d}{\delta t} + \frac{\delta (uA)}{\delta x} = 0 \qquad 1$$

$$\frac{\delta (Au)}{\delta t} + \frac{\delta (Au^2)}{\delta x} = -gA\frac{\delta S}{\delta x} - g\frac{Ad}{\rho}\frac{\delta \rho}{\delta x} - \frac{gAu|u|}{Rch^2} \qquad 2$$

$$\frac{\delta (Ac)}{\delta t} + \frac{\delta (Auc)}{\delta x} = \frac{\delta}{\delta x}(AK\frac{\delta c}{\delta x}) + I + L \qquad 3$$

The symbols are defined in the Appendix. Equations 1 and 2 have been written as explicit finite difference approximations and implemented in a computer programme to compute the varying current and depth distributions, along the estuary and in time, for given tidal range and river discharge. The friction coefficients were calibrated to match simulated conditions and measurements as closely as possible. Figure 3 shows the whole computer programme system that makes up the model. Section (a) solves the hydrodynamic equations while the programmes in section (b) are used to display and check the output. Comparisons between the predicted and observed currents for the survey on 10 July 1979 are shown on Figure 4. The model reproduces the progression of high water up the estuary. Also, the ratios between flood and ebb duration and the relative current amplitude are reproduced well for the stations where observations were made.

Equation 3 is also approximated in finite difference form and computerised. The solution of equation 3 uses the results from the hydrodynamic calculations. This section was calibrated against the measured salinity distribution and the coefficients K in the dispersion term varied to simulate the mixing conditions in the estuary. This term is an approximation to the flux of component c by dispersive action and includes the effects of several processes - turbulent mixing, shear effect, lateral and vertical circulations. Section (c) on Figure 3 relates to the mass balance solution and the programmes in (d) are used for examining the results, comparing with observed data and presenting the predictions in graphical form. When the mass balance section has been set up the dispersion coefficients are used in subsequent solutions of equation 3 for other parameters.

To compute dissolved oxygen (DO) levels in the estuary two solutions of equation 3 are required for each time step through the tide. First the level of biological oxygen demand (BOD) is computed accounting for the input from outfalls, the river and sea and the loss due to degradation in the previous time step. The DO distribution is obtained with a second solution of 3 using the loss due to BOD degradation and inputs from surface reaeration and river flow. A com-

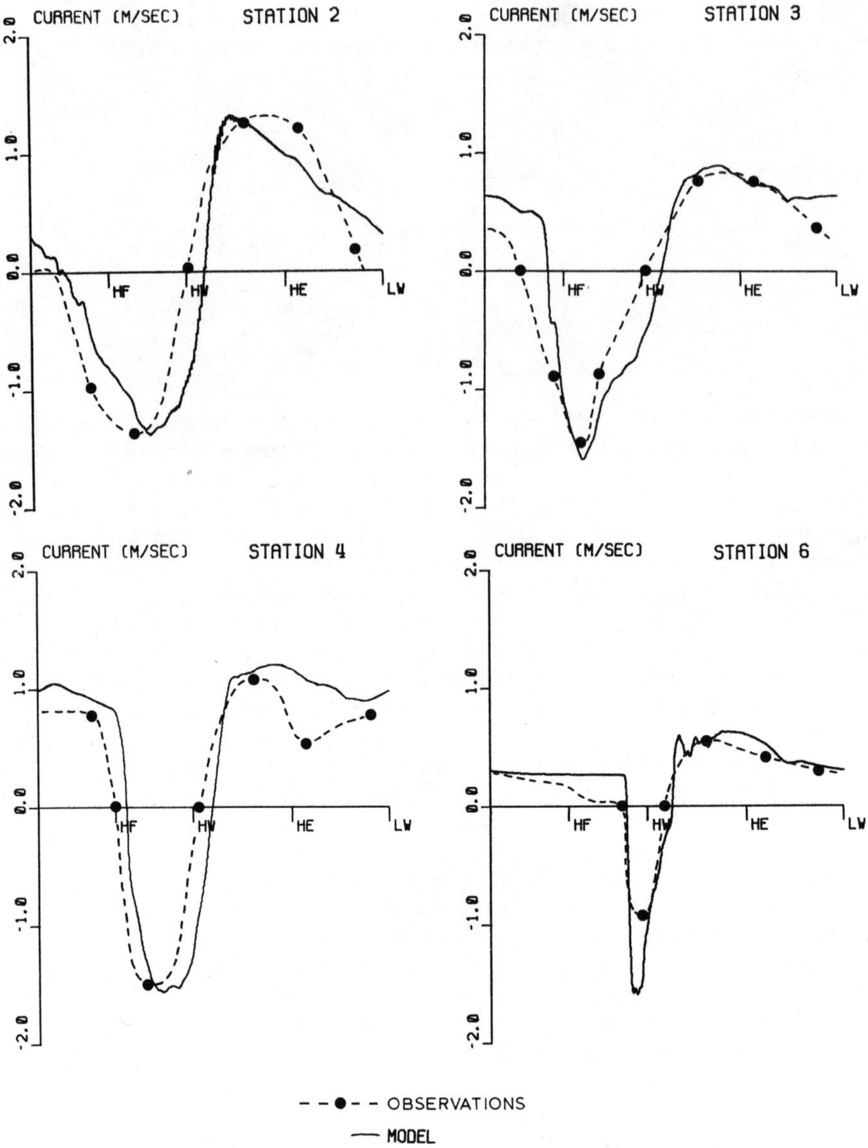


Fig. 4 Comparison of observed and predicted currents at stations 2(a), 3(b), 4(c) and 6(d) on 10.7.79. HF - half flood, HW - high water, HE - half ebb, LW - low water.

parison of the observed and predicted oxygen levels for the 1979 survey period is shown in Figure 5. The model has been used to assess the impact of the individual BOD discharges to the estuary and the mode of these discharges, and to determine the main causes for

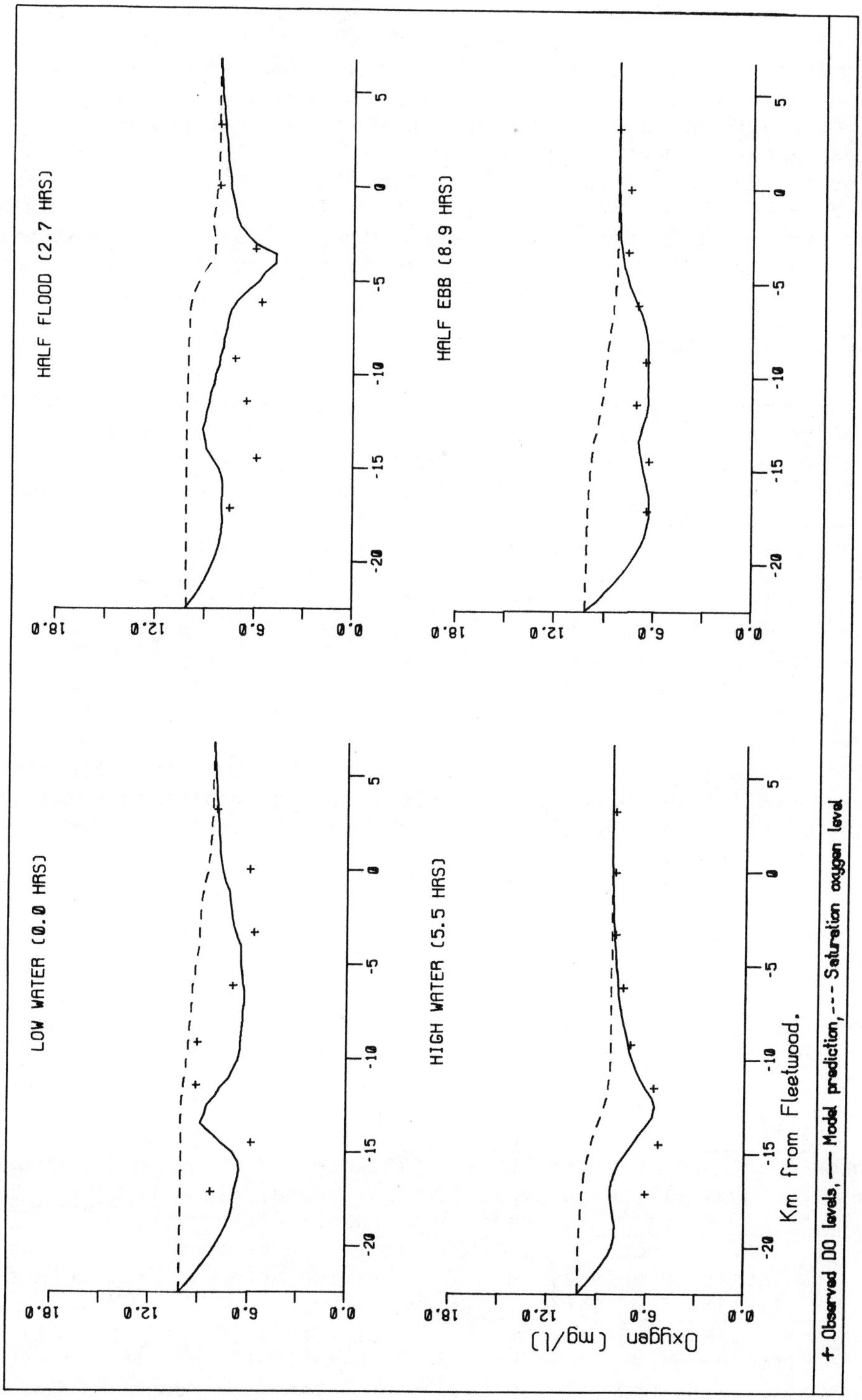


Fig. 5 Wyre estuary, one dimensional model. Oxygen prediction for 10.7.79.

lowered oxygen levels in the estuary. The predicted BOD levels for the upper three stations were much lower than measured values which has indicated the possibility of some form of BOD input in this area.

The model has also been used to look at other dissolved substances in the estuary such as ammonia. High ammonia levels (1.8 ppm) were found in the upper estuary but these were not reproduced by the model when using the known input loads for ammonia. This was observed over several surveys and some further investigation eventually brought to light an additional input of ammonia in the upper estuary.

## DISCUSSION AND CONCLUSIONS

The cross-sectionally averaged equations for flow and mass balance in an estuary are well established in the literature and have been given by many authors (e.g. Harleman, 1971, O'Connor and Thomason, 1971). Many models have been based on the numerical solution of some or all of these equations (e.g. Rossiter and Lennon, 1963, Thatcher and Harleman, 1972). The aim of this work has been to apply these methods to the Wyre estuary to quantify the effects of each of the effluents being discharged. The Wyre has a large intertidal volume and a small low water volume (the estuary is largely a river at low tide) requiring the model to handle a wide range of conditions. The resulting predictions show that the model can reproduce known conditions and can be a good tool for investigating possible changes in discharged loads and positions. At present the model predicts for a conservative substance or for BOD/DO interaction, but it has scope for enhancement to predict for a degrading or volatile substance and to include nitrification- denitrification in the oxygen prediction section.

## REFERENCES

Harleman D. R. F., 1971.One-dimensional models. In *Estuary Modelling; An assessment*. Tractor Inc, Austin Texas, Ch 3.

Inglis C. C., Kestner F. G. T., 1958.Long term effects of training walls, reclamation and dredging on estuaries. *Proceedings of the Institute of Civil Engineers*, A58, 6268, 193-215.

O'Connor D. J. and Thomason R. V., 1971.Water Quality Models; Chemical, Physical and Biological Constituents. In *Estuary Modelling; An Assessment*, Tracor Inc, Austin, Texas, Ch. 4.

Rossiter J. R., Lennon G. W., 1963.Computation of tidal conditions in the Thames estuary by the initial value method. *Proceedings of the Institute of Civil Engineers* 31, p. 25-57.

Thatcher M. L. and Harleman D. R. F., 1972.Prediction of unsteady salinity intrusion in estuaries- mathematical model and user's manual. *Massachusetts Institute of Technology School of Engineering Ralph M Parsons Laboratory*, Report No. 159.

## APPENDIX

Symbols;

| | |
|---|---|
| A(x,t) | cross-section area |
| b(x,t) | surface width of the estuary |
| c(x,t) | cross-section average concentration of a dissolved substance |
| Ch(x,t) | Chezy's friction coefficient |
| d(x,t) | depth of water |
| g | acceleration due to gravity |
| I(x,t) | input loads |
| K(x,t) | dispersion coefficient |
| L(x,t) | loss due to degradation |
| R(x,t) | hydraulic radius |
| t | time |
| u(x,t) | cross-section average current |
| x | distance from the head of the estuary |
| s(x,t) | surface elevation relative to the mean tide level |
| p(x,t) | density |

# DEVELOPMENT OF A MODEL OF BACTERIAL POLLUTION OF THE CLYDE ESTUARY

J.C. Curran and D.P. Milne

Clyde River Purification Board
East Kilbride, Glasgow, G75 OLA
Scotland

## INTRODUCTION

Criticisms have been made of the use of faecal coliforms (f.c.) as an indicator (Dutka, 1973; Bonde, 1975) of pathogenic bacterial or viral pollutin and of the use of a specific f.c. concentration threshold as an indication of human health hazard (Cabelli, 1981), yet it has become an accepted procedure (C.E.C., 1975) and is widely applied in estuarine and coastal pollution management.

Despite increasing reliance on microbiological monitoring (C.E.C., 1979), there remains no clear understanding of f.c. adaptability to the hostile marine environment (Gould and Munro, 1981) and therefore of the applicability of f.c. concentration predictions in areas affected by pollution sources.

A numerical model of f.c. concentrations in the Clyde Estuary is being developed, partly to provide an additional pollution management technique and partly to investigate f.c. environmental adaptability with the intention of improving existing and routinely-used, empirical, semi-empirical and theoretical models of pollution-field generation by proposed marine sewage outfalls (C.R.P.B., 1981).

## MODEL FORMULATION

Figure 1 illustrates the part of the Clyde Estuary for which the model is operative, the physical aspects being based exactly upon those devised for an existing, well-tested, finite-element water quality model (Curran, 1981).

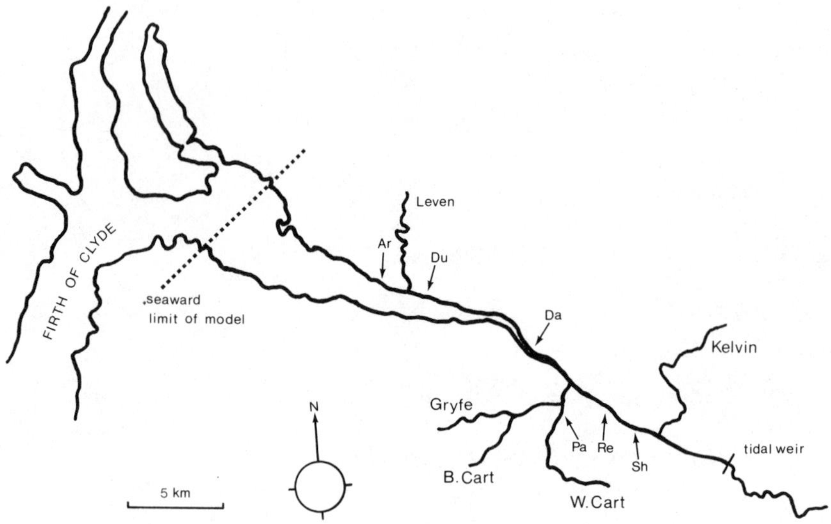


Fig. 1 Clyde Estuary and its major influents.
Sh: Shieldhall S.T.W. (approx. 2.0 cumecs, primary settlement)
Re: Renfrew S.T.W. (0.2 cumecs, primary)
Pa: Paisley S.T.W. (0.7 cumecs, secondary)
Da: Dalmuir S.T.W. (2.5 cumecs, primary)
Du: Dumbarton S.T.W. (0.1 cumecs, primary)
Ar: Ardoch/Havock S.T.W. (0.15 cumecs, primary)

The tidal-averaged, one-dimensional form of the convection-diffusion equation (Gilligan, 1972) is applied to suspended solids and f.c. concentrations in the estuary.

$$\frac{\delta}{\delta x}(QL - EA\frac{\delta L}{\delta x}) = m_L - k_L AL \qquad 1$$

$$\frac{\delta}{\delta x}(Qc - EA\frac{\delta c}{\delta x}) = m_c - k_c Ac \qquad 2$$

where the dispersion coefficient is defined as

$$E = Qs/A\frac{ds}{dx} \qquad 3$$

and has the convenient modelling form displayed in Figure 2.

The mathematical treatment of polluting influents, cross-sectional areas, tidal excursions, temperature, salinity data and Galerkin weighting functions is exactly as for the water quality model (Curran, 1981) and, once again, only first-order reaction rates are employed.

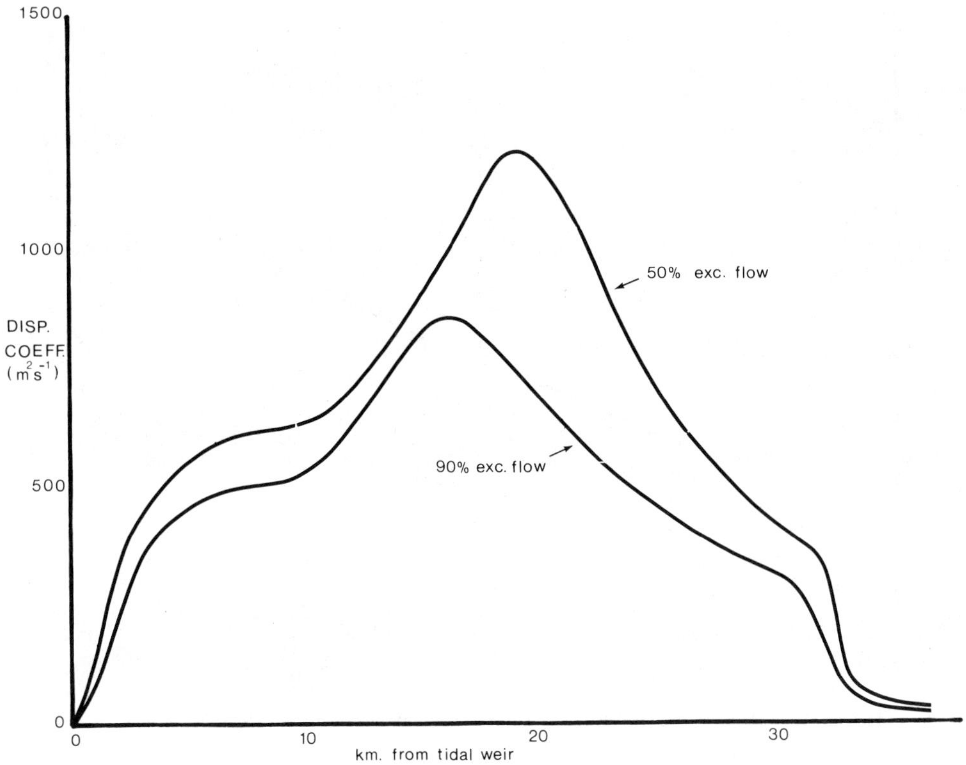


Fig. 2 Dispersion coefficient, E, at median and low flows.

## REACTION RATES

### (1) Suspended solids

A first-order reaction rate has been successfully used (Uchrin and Weber, 1981) to model settlement of suspended solids and it has been shown (Dyer, 1979) that the deposition rate is proportional to $(1 - \Gamma / \Gamma_c)$ Integration of an assumed logarithmic velocity profile in the 8m deep dredged channel and the assumption of representative values of $\Gamma_c$ and of dynamic roughness length (Dyer, 1979) provide the expression.

$$\text{rate} = \frac{-dL/dt}{L} = k_L = k_{ss}(1 - 40(\tfrac{Q}{A}+U)^2), \quad k_L \nless 0 \qquad 4$$

### (2) Faecal Coliforms

It has been found (Canale et al, 1973; Gould and Munro, 1981) that first-order reaction rates adequately describe f.c. die-off, mortality or disappearance.

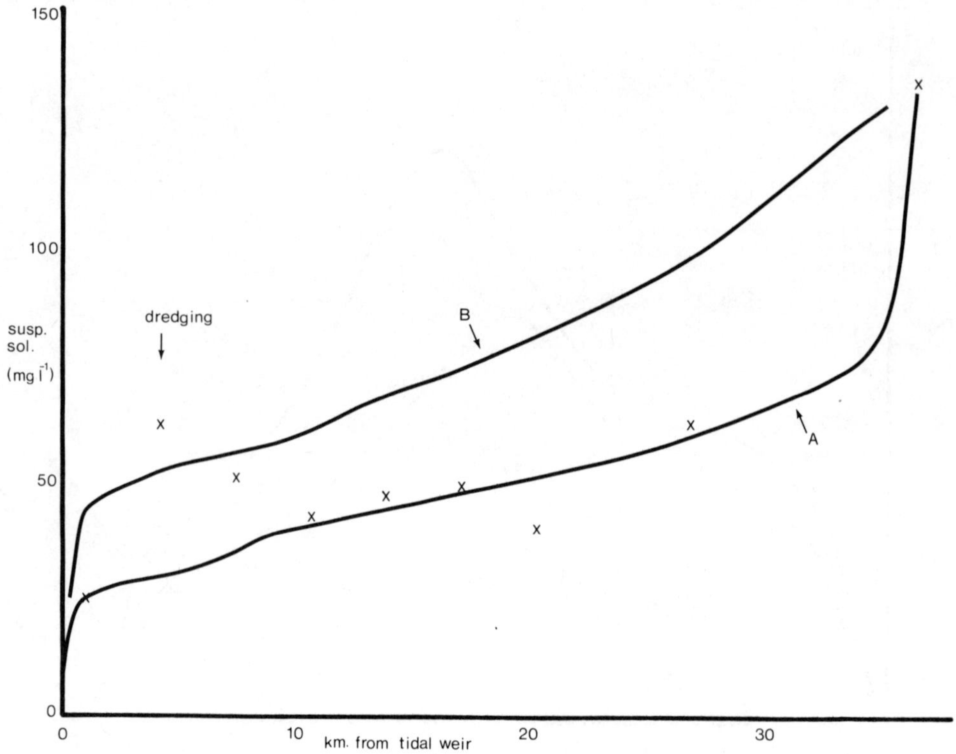


Fig. 3 A: model prediction of suspended solids on 27/4/82 x denotes depth-averaged survey results (> 1.5 μm)
B: model prediction with zero deposition coefficient, $k_L = 0$

(a) Effect of Sunlight: The mortality induced by solar irradiation may be expressed (Mitchell and Chamberlin, 1975) as

$$\text{die-off} = -\frac{dc/dt}{c} = k_I I e^{-\alpha z} \qquad 5$$

while integration with depth and the assumption that $\alpha = 0.5L$ (Dyer, 1979) implies that

$$\text{die-off} = k_f f M 10^3/L \qquad 6$$

where f is determined from a nomogram (Lumb, 1964) relating the fraction of potential solar irradiation to cloud cover and type.

(b) Physico-chemical effects: Die-off in the dark increases proportionately and by less than one order of magnitude (Gould and Munro, 1981) across the range of salinities encountered in an estuary and when combined with a representative temperature dependence (Mancini, 1978) then

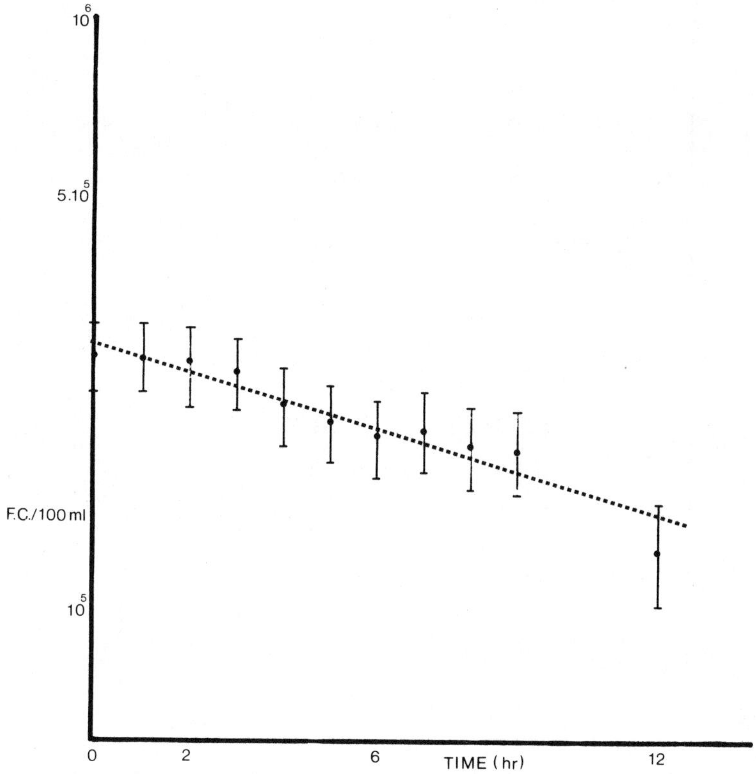


Fig. 4 Faecal coliform concentrations determined under agitated conditions. Temp. = 12°C, salinity = 14.0‰, susp. sol. = 46 mgl$^{-1}$ 95% confidence limits are shown (H.M.S.O., 1969)

$$\text{die-off} = k_s(1+0.2s)\times 1.07^{(T-20)} \qquad 7$$

(c) Deposition: It has been suggested (Mitchell and Chamberlin, 1975) that direct deposition of f.c. which are adsorbed onto suspended solids may be of significance, hence

$$\text{disappearance} = k_d.k_L \qquad 8$$

where $k_L$ is described by equation (4).

The combined disappearance rate, $k_c$, of f.c. in the estuarine environment is taken as the sum of the terms in equations (6) to (8). Regeneration, predation, environmental toxicity and bacteriophage activity (Mitchell and Chamberlin, 1975) are not explicitly modeled but may be included implicitly in the coefficients $k_f$, $k_s$, and $k_d$.

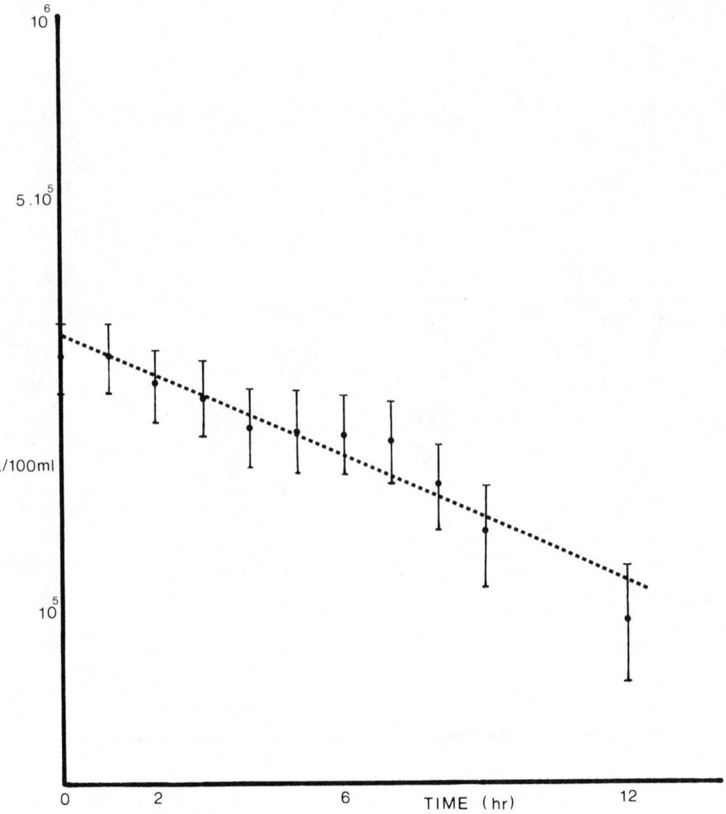


Fig. 5 Faecal coliform concentrations determined under quiescent conditions.

## CALIBRATION OF MODEL

### Survey Data

After a dry spell of weather, on 27th April, 1982, a survey of suspended solids (fraction > 1.5 μm) was made at eight stations and at three depths along the length of the estuary and f.c. concentrations at seven stations and three depths; routine microbiological methods (C.R.P.B., 1982) were employed with all pre-secondary incubation stages being performed on board the survey vessel. The seaward boundary condition was similarly determined and the major influents of Figure 1 simultaneously sampled. Routine gauging by the Clyde River Purification Board provides four-day mean flows for the inflowing rivers for which the percentage exceedences were found to lie between 88% and 95%; low and consequently steady flows are necessary in order that a steady-state be approached in the estuary for which the retention period may be lengthy (Mackay and Gilligan, 1972).

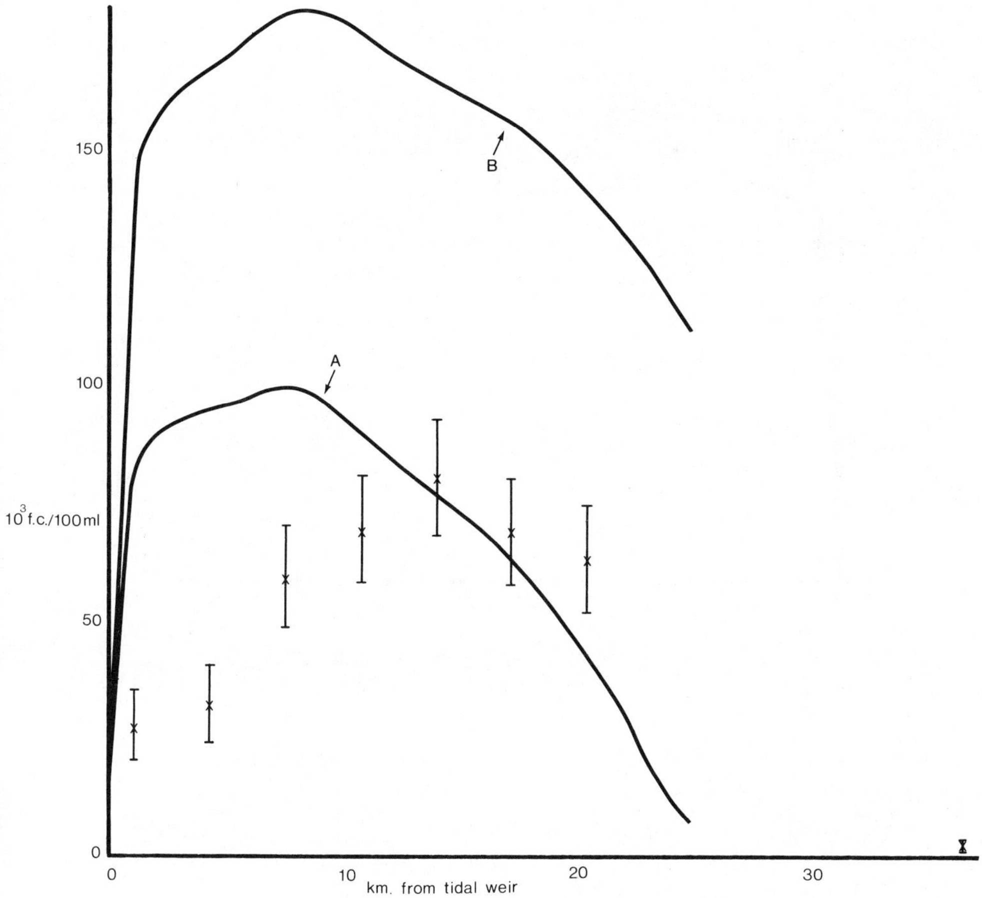


Fig. 6 A: model prediction of faecal coliforms on 27/4/82 x denotes depth-averaged survey results 95% confidence interval.
B: model prediction with zero disappearance rate, $k_f=k_s=k_d=0$.

Four-day mean flows for the three major sewage treatment works were provided by the Strathclyde Regional Analyst and other discharges were estimated from population figures (Bartlett, 1970).

## Operation of Model

The model is written in Data General extended Algol and, with 139 nodes, it occupies 64 kbytes of memory and consumes 4.5 minutes CPU time.

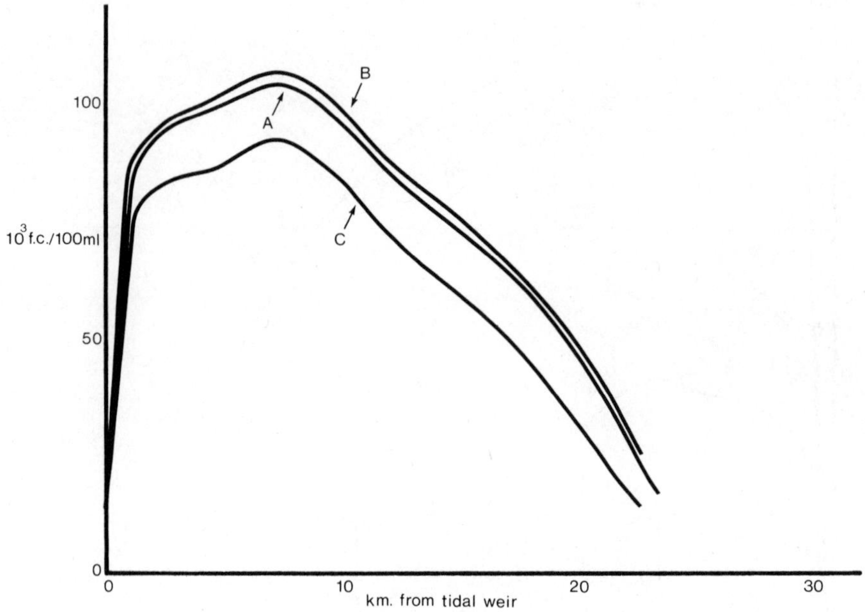


Fig. 7 A: calibrated model for 27/4/82
B: estuarine water temp. = 5°C
C: estuarine water temp. = 20°C

## Suspended Solids

Figure 3 reveals good agreement between the model and the survey data when run with a numerical value of $k_{ss}$ of $5 \times 1^{-6}$.

## Die-off Experiment

17 km seawards of the tidal weir, samples of estuary surface water were collected in sterile, light-excluding bottles; subsamples for the determination of temperature, salinity and suspended solids were taken concurrently. After one on-site f.c. determination, two bottles were incubated in the laboratory at ambient estuary temperature and in complete darkness for 75 hours; one bottle left totally undisturbed and the other thoroughly shaken before extraction of duplicate aliquots from each for microbiological analysis by the standard membrane filtration method (H.M.S.O., 1969) with the incubation regime modified to 44°C for 18 hours. The results are displayed in Figures 4 and 5.

Figure 4 indicates a physico-chemical mortality rate for the first 12 hours of approximately 0.06 $hr^{-1}$ ($T_{90} \approx$ 40 hr) which agrees reasonably well with other published data (Mancini, 1978; Matsumoto and Omura, 1980), while the difference between the rates of Figures 4 and 5 suggests a disappearance rate due to deposition of approximately 0.015 $hr^{-1}$ ($T_{90} \simeq$ 150 hr).

These results imply that $k_s$ and $k_d$ of equations (7) and (8) must be ascribed the numerical values of $7.5 \times 10^{16}$ and 0.85 respectively.

## Faecal Coliforms

The coefficient $k_f$ remains as the adjustable calibration parameter of the model; Figure 6 illustrates the fit between the model predictions and the survey results using a numerical value of $k_f$ of $6 \times 10^{-5}$.

Figure 7 indicates the relatively minor effect of the annual range of estuarine water temperature on predicted f.c. concentrations while Figure 8 reveals the overwhelming effect of sunlight intensity changes even at one time of year.

## RESULTS

The calibrated coefficients for 27 April 1982 may be used in the appropriate equations (6) to (8) in addition to the relevant survey data on suspended solids, cloud cover and type, salinity and temperature to determine the individual disappearance rates experienced at a point near the upper end of the estuary; Table 1 presents the results.

The total disappearance rate is approximately 1.2 $hr^{-1}$ ($T_{90} \simeq$ 2.0 hr) and the maximum range for a cloudy midwinter day to a cloudless midsummer day might be expected to be approximately 0.07 $hr^{-1}$ to 2.6 $hr^{-1}$ ($T_{90} \simeq$ 33 hr to 0.9 hr). These figures are consistent with other published data (Gameson and Gould, 1975; Irving, 1977; Gould and Munro, 1981), although a great range of results is found throughout the relevant literature.

## CONCLUSIONS

It has been suggested (Mitchell and Chamberlin, 1975) that numerical modelling may be employed usefully as an alternative to the many, and often criticised (Haremoes, 1970; Zanoni and Fleissner, 1982), experiments which have been devised to determine f.c. die-off rates.

This model is simple and requires considerable further improvement, validation and testing against survey data, yet its initial results appear realistic and acceptable. The model's relatively poor

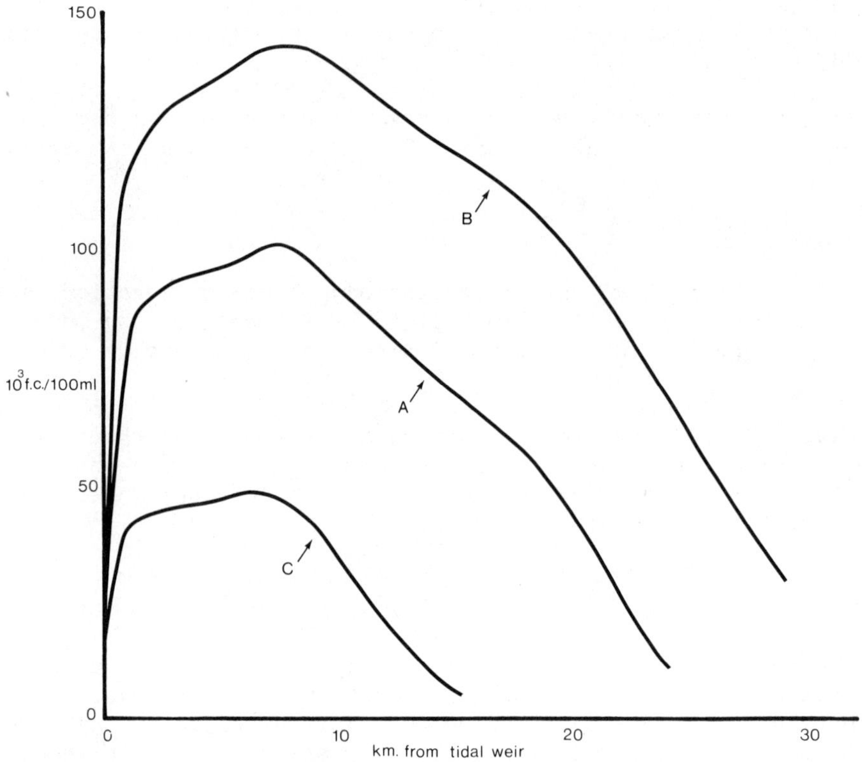


Fig. 8 A: calibrated model for 27/4/82
B: assumed total cloud cover for 27/4/82
C: assumed zero cloud cover for 27/4/82

performance at the upper end of the estuary may be due to a atypically low value of $k_d$ (see Table 1) derived from the die-off experiment; the latter will be repeated in modified form.

Although bacterial pollution of the Clyde Estuary is clearly severe, and with limited prospect of early improvement, the model may be beneficial in investigations of the effect, for example, of freshwater abstraction from the headwaters, of increased sewage loadings or of the impending improvement of effluent quality from the Shieldhall sewage treatment works (Greer, 1981).

Additionally, if the calibration values of die-off are realistic then they may be helpful in the more general assessment or prediction of pollution spread from proposed marine outfalls and therefore may aid in the establishment of design requirements for discharge. Significant die-off rates may be included advantageously in calculations based on dye patch dilution rates and on current meter statistics which are particularly appropriate for the derivation of percentage

Table 1 Calculated disappearance rates, 4 km seawards of tidal weir on 27/4/82

| Cause of disappearance | Disappearance rate (base e), $hr^{-1}$ | $T_{90}$, hr |
|---|---|---|
| deposition | 0.012 | 185 |
| physico-chemical | 0.04 | 55 |
| sunlight | 1.1 | 2.1 |

exceedence values of f.c. concentration for comparison with accepted standards (C.E.C., 1975); this approach is particularly valuable if retrospective validation of the consent-determination survey work and analysis is to be effected by routine, post-commission, shore-line microbiological surveys. Representative die-off may also be included in simple deterministic models of effluent plume development, based on equations (Brooks, 1960) which use either assumed or locally measured diffusion coefficients in order to predict downstream isopleths of f.c. concentration in addition to the more usual (C.R.P.B., 1981) limits of smell or visual nuisance (Newton, 1973).

NOTATION

- $x$ distance along estuary from tidal weir (m)
- $Q$ freshwater flow ($m^3s^{-1}$)
- $E$ dispersion coefficient, defined by equation (3) ($m^2s^{-1}$)
- $A$ cross-sectional area of estuary ($m^2$)
- $L$ suspended solids concentration ($mgl^{-1}$)
- $c$ faecal coliform concentration (f.c./100 ml)
- $m$ input pr unit length of estuary
- $k$ first-order reaction rate (base e) or die-off/ mortality/ disappearance rate ($s^{-1}$)
- $s$ salinity (%)
- $\Gamma$ bed shear stress
- $\Gamma c$ critical value of $\Gamma$ below which deposition occurs
- $U$ mean tidal flow = tidal excursion/0.5x tidal period ($ms^{-1}$)
- $I$ solar radiation intensity
- $\alpha$ attenuation coefficient
- $Z$ depth
- $f$ fraction of potential solar irradiation
- $M$ seasonality factor dependent upon month of the year = sine of solar elevation
- $T$ temperature (°C)
- $T_{90}$ time for 90% disappearance or 10% survival

Subscripts L and ss, I and f, s, d, c refer to suspended solids, radiation, physico-chemical effects, deposition, faecal coliforms respectively.

## ACKNOWLEDGEMENTS

This paper is published by the permission of Mr. D. Hammerton, Director of the Clyde River Purification Board. Sewage works' effluent data were provided by the Department of the Regional Chemist, Public Analyst and Agricultural Analyst with the permission of the Director of Sewerage, Strathclyde Regional Council. The encouragement and advice of Dr. A.J. Newton and the computing assistance of Mr. J.M. Rooney are gratefully acknowledged.

## REFERENCES

Bartlett, R.E. 1970. *Public Health Engineering - Design in Metric; sewerage*. Elsevier, 208 pp.

Bonde, G.J. 1975. Bacterial indicators of sewage pollution. In *Discharge of Sewage from Sea Outfalls*. Pergamon Press

Brooks, N.H. 1960, Diffusion of sewage effluent in any ocean current. In *Waste Disposal in the Marine Environment*. Pergamon Press.

Cabelli, V.J. 1981. A health effect data base for the derivation of microbial guidelines for municipal sewage effluents. In *Coastal Discharges*. Institute Civil Engineering, London.

Canale, R.P., Patterson, R.L., Gannon, J.J. and Powers, W.F. 1973. Water quality models for total coliform. *Journal of the Water Pollution Control Federation* 45, 325-336.

Commission of the European Communities 1976, Council directive concerning the quality of bathing water. *Official J. of the Eur. Comm. No. L31*.

Commission of the European Communities 1979, Council directive on the quality required of shellfish waters. *Official J. of the Eur. Comm. No. L281*.

Clyde River Purification Board 1981. A report on the potential of the adjacent tidal waters to accept domestic sewage disposal from a proposed development on Rosneath Peninsula. *Technical Report* No. 63

Clyde River Purification Board 1982. The microbiological quality of bathing beaches and nearshore waters in the CRPB's area. 1977-80. *Technical Report* No. 66

Curran, J.C. 1981. A finite-element model of pollution in the Clyde Estuary: formulation, validation, utilisation. *Applied Mathematical Modelling*, 5, 137-142.

Dutka, B.J. 1973. Coliforms are an inadequate index of water quality. *Journal Environmental Health*, 36, 39-46.

Dyer, K.R. 1979. *Estuarine hydrography and sedimentation*. Cambridge Univ. Press., 230 pp.

Gameson, A.L.H. and Gould, D.J. 1975.Effects of solar radiation on the mortality of some terrestrial bacteria. In Discharge of Sewage from Sea Outfalls. Pergamon Press.

Gilligan, R.M. 1972.Forecasting the effects of polluting discharges on estuaries. Water Pollution Research Laboratory, Reprint No. 685.

Gould, D.J. and Munro, D. 1981.Relevance of microbial mortality to outfall design. In Coastal Discharges. Institute Civil Engineering

Greer, W.T. 1981.Design and construction of Shieldhall seqage treatment works, Glasgow. Water Pollution Control, 80, 260-272.

Harremoes, P. 1970.Field determination of bacterial disappearance in seawater. Water Research, 4, 737-749.

H.M.S.O., 1969,The bacteriological examination of water supplies. Report No. 71.

Irving, T.E. 1977.Preliminary investigations on the effects of prticulate matter on the fate of sewage bacteria in the sea. Water Research Centre, Technical Report 55.

Lumb, F.E. 1964.The influence of cloud on hourly amounts of total solar radiation at the sea surface. Quarterly Journal Royal Meteorological Society, 90, 43-56.

Mackay, D.W. and Gilligan, J. 1972.The relative importance of freshwater input, temperature and tidal range in determining levels of dissolved oxygen in a polluted estuary. Water Research, 6, 183-190.

Mancini, J.L. 1978.Numerical estimates of coliform mortality rates under various conditions. Journal of the Water Pollution Control Federation, 50, 2477-2484.

Matsumoto, J. and Omura, T. 1980.Some factors affecting the survival of faecal indicator bacteria in seawater. Technology Report, Tohoku Univ., 45, 169-185.

Mitchell, R. and Chamberlin, C. 1975.Factors influencing the survival of enteric microorganisms in the sea: an overview. In Discharge of Sewage from Sea Outfalls. Pergamon Press.

Newton, J.R. 1975.Factors affecting slick formation at marine sewage outfalls. In Pollution Criteria for Estuaries. Pentech Press.

Uchrin, C.G. and Weber, W.J. 1981.Modelling suspended solids and bacteria in Ford Lake. Journal Environmental Engineering Division, ASCE, 107, 975-993.

Zanoni, A.E. and Fleissner, J.T. 1982.Indicator bacteria survival under laboratory conditions. Journal of the Water Pollution Control Federation, 54, 500-503.

# THE EFFECT OF PARTICULATE BOD ON THE OXYGEN BALANCE OF A MUDDY ESTUARY

Jacqueline M. Maskell

Tidal Computations Group
Hydraulics Research Ltd.
Wallingford, OC10 8BA
England

## ABSTRACT

A recurrent phenomenon in the very turbid tributary estuaries of the Severn is a larger depletion of dissolved oxygen on spring tides compared with neap tides. This is the reverse of what usually occurs in estuaries with low suspended mud concentrations. This paper describes the results of mathematical model simulations of the suspended mud transport and oxygen balance in the Parrett Estuary and shows that naturally occurring particulate BOD (slowly degrading detritus), which is part of the suspended mud load, is almost certainly responsible for the phenomenon.

## INTRODUCTION

A striking phenomenon in turbid estuaries in the U.K. is the larger depletion of dissolved oxygen on spring tides compared with neap tides. This is the reverse of what occurs in estuaries with low suspended mud concentrations where the dominant process is dilution of effluent by tidal water which is usually significantly less on neap tides.

Observations in the Usk Estuary during 1976-1980 showed a typical oxygen sag on spring tides to have minimum dissolved oxygen concentrations of 1.5 mg/l whereas the typical minimum value on neap tides was 3 mg/l. In the Wye Estuary, which has no significant polluting loads, dissolved oxygen levels as low as 60% of the saturation value occur during spring tides (Welsh Water Authority, 1981).

Many U.K. water authorities have found it impossible to calibrate their oxygen balance models on both neap and spring tides without adding a extra oxygen demand, which is often assumed to be proportional to the observed concentration of suspended solids. However, it is generally not known what proportion of resuspended particulate BOD is derived from crude sewage discharges or from natural marine detritus.

In 1978, the author embarked on the development of a one-dimensional, cross-sectionally averaged, mathematical model of oxygen balance in an estuary which simulated the intertidal movement of water, salt, mud and dissolved and particulate BOD. The work was funded by the Department of the Environment and was done with advice from the Water Research Centre at Stevenage.

The first application of the model was to simulate conditions in the Parrett Estuary for Wessex Water Authority, who commissioned the Hydraulics Research Station to predict the effect of a proposed barrage near Bridgewater on the oxygen balance in this very turbid estuary as part of a feasibility study. Although the model study met its main objectives the model was unable to reproduce the correct oxygen sag on spring tides for existing conditions.

This paper describes a series of numerical experiments carried out with the hydraulically proved Parrett model during the period 1981-83. The tests were designed to investigate the effect of different forms of particulate and dissolved BOD on the oxygen balance of the estuary.

## The Parrett Estuary

The Parrett Estuary was modelled for a distance of about 50 km from Hinkley Point, in the seaward approaches, to the tidal limits as shown in Fig. 1. For the purposes of the model the estuary was divided into fifty five elements each 1 km long.

The cross-sectional area of the channel decreases rapidly and the bed level rises in the landward direction. The bed of Bridgewater Bay consists of soft easily eroded mud. The bed surface sediment between Stert and Dunball is predominantly fine sand, upstream of Dunball the bed surface sediment becomes progressively finer.

Hydraulic data for proving the model was obtained from field surveys made by the Hydraulics Research Station during a spring and a neap tide in October 1977. There was only a very limited amount of up-to-date water quality data available for the estuary which consisted of a few dissolved oxygen observations made in October 1977.

A detailed description of the model is beyond the scope of this paper but is given in a Hydraulics Research report (Maskell, 1984).

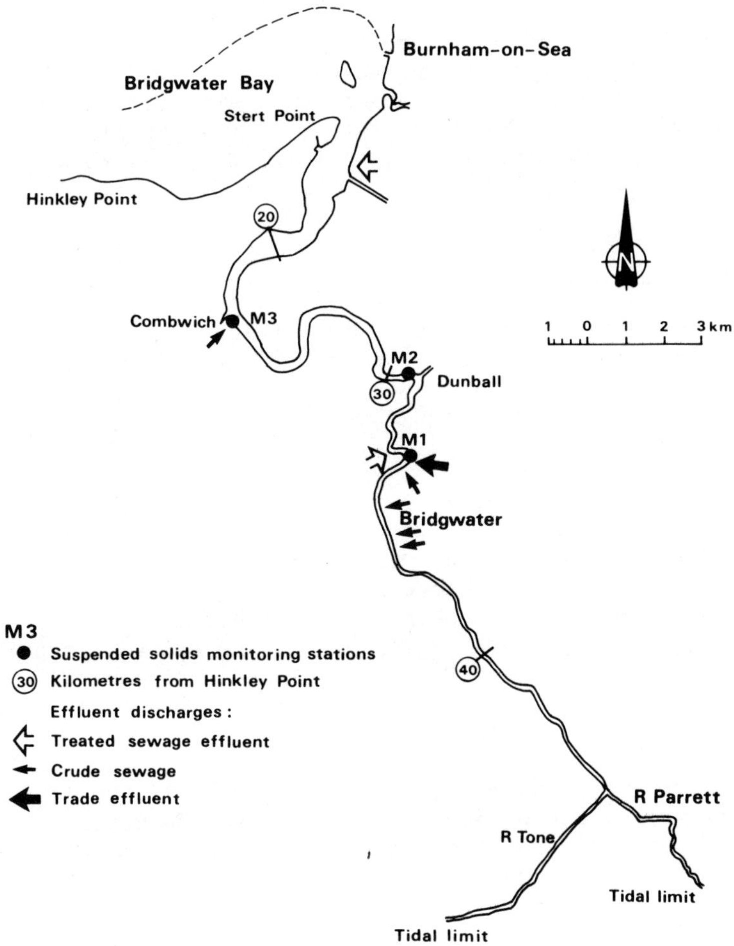


Fig. 1 Plan of the Parrett estuary

It is sufficient to say that the model was calibrated so that it satisfactorily simulated the following tidal processes in the estuary during a period of steady low fluvial flows:

- tidal propagation
- saline intrusion
- mud transport

The estuary has a strongly semi-diurnal tide with a spring tide range of 13 m at Hinkley Point, one of the largest tidal ranges in the U.K. The depth of water in the lower and middle estuary at low water springs is about one metre so that there is a large tidal range to depth ratio. The volume of water contained in the estuary at low water is only a few percent of that at high water. The model was extended 16 km seaward of Stert Point so that it would contain the tidal volume of the estuary at low water. The estuary is cross-sectionally well mixed during periods of low fluvial flows. A comparison of salinity profiles at high and low water slack during a spring tide, when fluvial discharges totalled 3.7 cumecs, indicated that the tidal excursion is about 16 km during spring tides.

## Mud transport in the estuary

There are several time-scales and cycles of mud transport in the estuary. The first cycle is the seasonal movement of mud to and from the upper reaches of the tidal compartment depending on the variation of the fluvial discharges throughout the year.

The second cycle of mud transport occurs in the spring-neap cycle of 15 days as illustrated in Fig. 2. During neap tides the tidal currents are not strong enough to scour and carry into the estuary significant quantities of mud. Both the scouring and transporting capacity of the tidal flow increase rapidly with increasing tidal range as shown in Fig. 3. During spring tides about a hundred thousand tonnes of mud move in and out of the estuary past Stert Point. This process is modified by the interaction of tidal and fluvial flows, however the effect of fluvial flows becomes progressively less in the seaward reaches of the estuary.

The third important cycle of mud transport in the estuary is the semi-diurnal cycle during spring tides. Large quantities of mud are scoured into suspension by the strong tidal currents (Fig. 4). Most of the mud comes from the re-erosion of slack water and neap tide deposits on the bed of the seaward approaches to the estuary (seaward of Stert Point). The area-average concentrations are generally high enough to cause interference between the mud flocs and to give rise to a state of hindered settling, in which the settling velocity decreases as the concentration increases. During the high water slack period the mud in suspension settles to form a layer of fluid mud in the deeper parts of the channel and which has a well-defined horizontal interface with the clear water above. The dry density of these slack water layers is of the order of 0.1 tonnes $m^{-3}$ dry matter and their shear strength is about $0.1 N/m^2$. When the tide turns there soon becomes a point when the stress in the fluid mud exceeds its Bingham shear strength causing the whole mass to go into suspension in a short time. The model simulated the formation, dewatering and resuspension of fluid mud in the estuary.

The oxygen balance calculation, which was coupled interactively with the flow, salt and mud transport sections of the model using the same short timestep of 120 seconds, was based on theories developed by the Water Research Centre and modified to take account of particu-

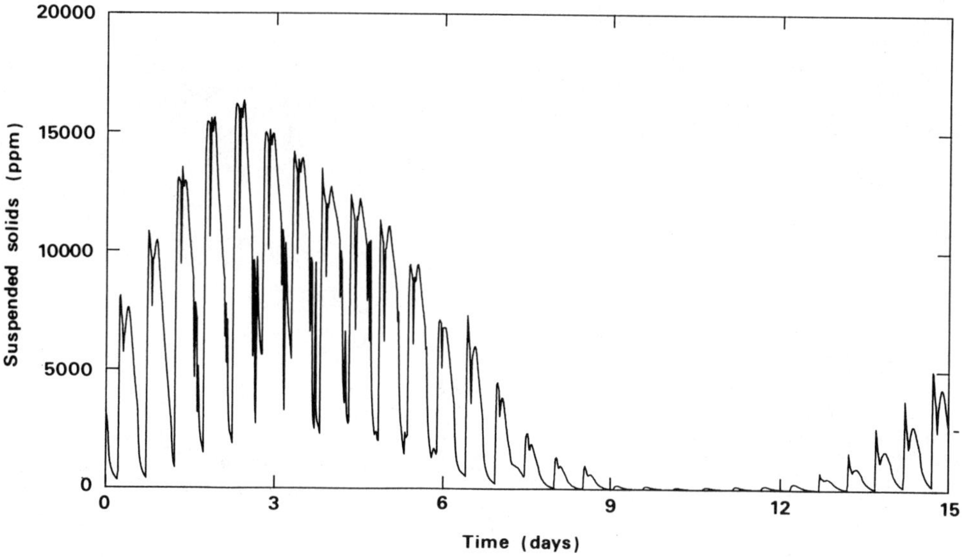


Fig. 2 Computed variation of suspended mud concentrations at Bridgewater

late BOD. The model calculated the effect of nine substances; fast and slow dissolved carbonaceous material, fast and slow particulate carbonaceous material, fast and slow organic nitrogen, ammoniacal nitrogen, nitrate and dissolved oxygen. It was assumed that both fractions of the particulate BOD would be wholly mixed with the marine mud, moving into and out of suspension accordingly. The model took separate account of pollutants trapped in the interstitial water in mud deposited on the bed.

## Model simulations

In order to investigate the role of particulate BOD on the oxygen balance of the estuary, the model was set up to simulate three conditions.

Firstly, the model was used to simulate conditions in October 1977 using the best estimates of the loads of dissolved and prticulate BOD for each of the effluent outfalls. The major outfalls, including an industrial effluent with a high particulate content, had been sampled and analysed by Wessex Water Authority in filtered and unfiltered states to determine the magnitude and oxidation rates of the dissolved and particulate fractions of BOD. The model was run for several repeating spring-neap tidal cycles until it reached a state of dynamic equilibrium with a repetition of the results every 15 days. The model results are illustrated in terms of the interti-

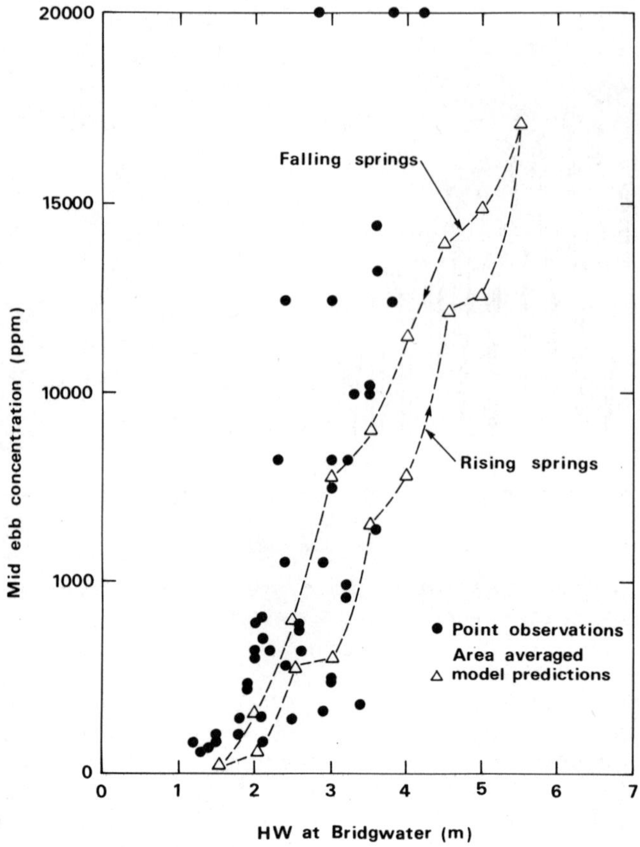


Fig. 3 Comparison of computed and observed mud concentrations at Cambwich

dal variation of dissolved oxygen levels over a spring neap cycle at Bridgewater in Fig. 5 and as a longitudinal profile early in the ebb period of a spring tide in Fig. 6. The model predictions overestimated the observed DO levels of about 40% saturation on spring tides. Fig. 5 shows higher levels of DO during spring tides when dilution is greatest. There appeared to be a major input of pollution missing from the model, one wich had a greater effect on spring tides than on neaps. The only source of such pollution seemed to be the marine mud which would contain natural detritus (dead phytoplankton, disintegrated faecal pellets, etc).

It was therefore decided to run the model for a second test which included a very slowly degrading particulate fraction of BOD which was input with the mud at the seaward boundary of the model on the flood tide. The rate of oxidation of slow particulate BOD in the sewage effluents had been taken as 0.04 $day^{-1}$, one fifth the standard

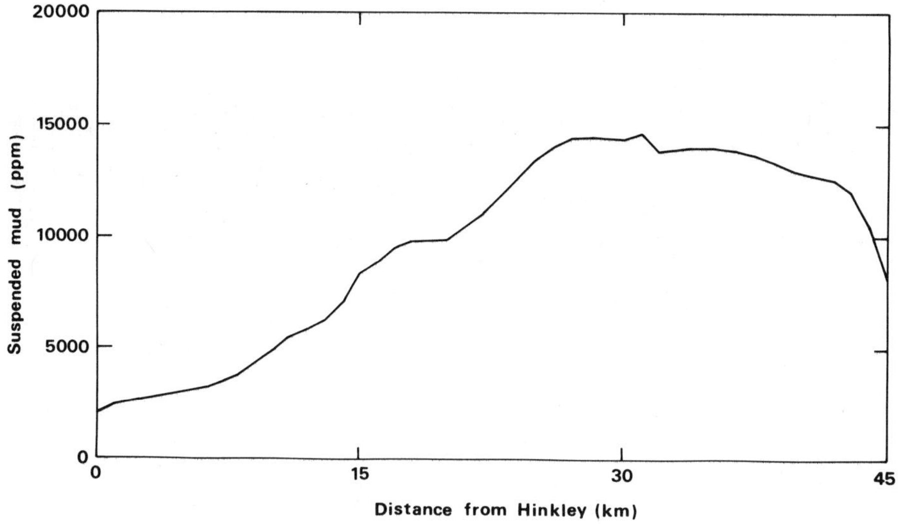


Fig. 4 Computed longitudinal variation of suspended mud concentration

rate. This was considered too fast for the slowly degrading detritus. Analysis of samples of bed deposits from the Parrett undertaken at HR indicated that the rate of oxidation was very slow but the small number of samples and the variation in results precluded any firm conclusion concerning the oxidation rate. All samples however showed a loss on ignition of between 8% and 10%. It is known that this is an overestimate of the actual carbon content due to the evaporation of water locked in the mud matrix but it was felt that a conservative estimate of the carbon content of 2% was realistic, giving an ultimate oxygen uptake of approximately 6% of the mass of suspended solids. The oxidation rate of the detritus was taken to be ten times slower than that normally used for the slow component of dissolved BOD, i.e. a value of 0.004 $day^{-1}$ was used. This agrees with findings on the degradation of sludge (Barrett et al, 1972).

The effect of the oxygen demand of the detritus is illustrated in Figs.5 and 6. These show that the effect of the detritus is very large on spring tides and negligible on neap tides, which reflects the variation in suspended mud concentration. The pattern of DO levels through the spring neap cycle agrees with what has been observed in the Usk Estuary. The minimum level of dissolved oxygen on spring tides is close to the reported values in the Parrett of 40% saturation. The position of the DO sag, early on the ebb of a spring tide, is shown in Fig. 6 and is displaced 15 km seaward compared to the first model simulation. The analysis of the longitudinal distribution of the oxygen demand due to natural detritus, dissolved and par-

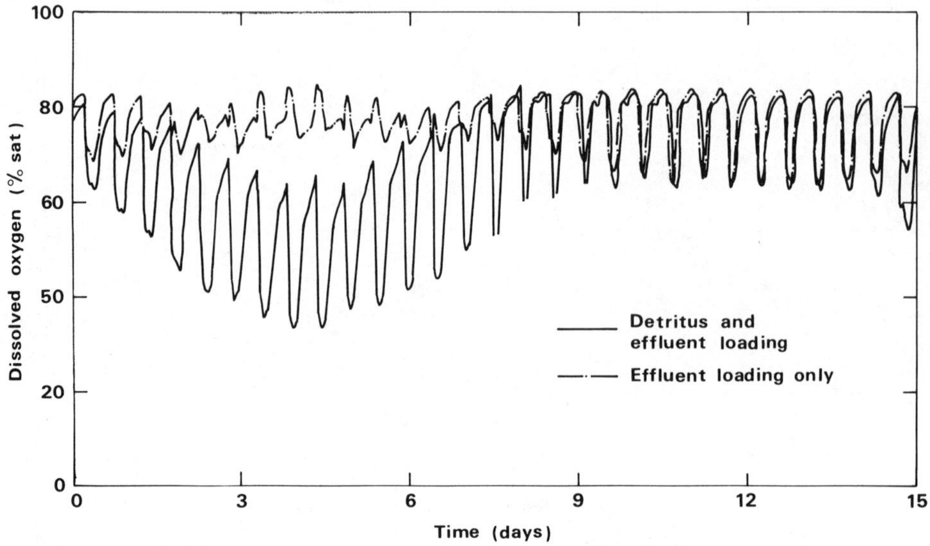


Fig. 5 Computed variation of dissolved oxygen at Bridgewater

ticulate BOD is shown in Fig. 7. It is apparent that the instantaneous oxygen demand due to natural detritus is significant compared with the other sources and reflects the longitudinal distribution of suspended mud concentration at the same time (Fig. 4).

The model was run for a third test with no artificial loads into the estuary. The results, illustrated in Fig. 6, show a minimum oxygen level of about 60% saturation in the middle of the estuary during a spring tide. This is similar to the reported oxygen levels in the Wye which has no significant artificial polluting loads.

The results of this final test indicate that there is a natural limit to which one can improve the dissolved oxygen level in a turbid estuary by reducing artificial loads.

## DISCUSSION AND CONCLUSIONS

The biochemical oxygen demand of naturally occurring marine detritus in the Parrett Estuary (approximately 2 mg/1/day), which is periodically resuspended during spring tides, is almost certainly greater than the BOD due to dissolved and particulate effluents discharged into the estuary.

The detritus is derived from the ecosystem of the Severn Estuary and a major source is thought to be dead phytoplankton from the more

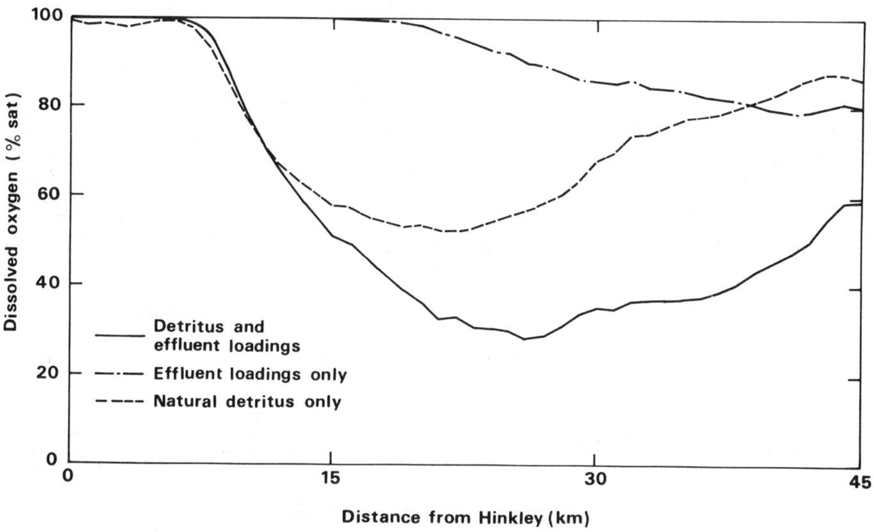


Fig. 6 Computed longitudinal variation of dissolved oxygen

turbid parts of the estuary. A more detailed study of the form and amount of the detritus produced could be made by using the GEM-BASE model developed at the Institute for Marine Environmental Research at Plymouth (Radford, 1981).

The observed levels of dissolved oxygen in the Parrett were simulated by assuming that the detritus had a rate constant for oxidation of about 0.004 $day^{-1}$. It was also assumed that the detritus accounted for about a quarter of the observed 10% loss on ignition of the marine mud.

The particulate fraction of the BOD in the effluents discharged into the middle of the estuary, which has a rate constant of 0.2 $day^{-1}$, tends to be transported upstream and to draw down dissolved oxygen levels in the muddy silting zone at the head of the estuary. The dissolved fraction of the BOD in the effluents tends to exert its load in the vicinity of the main outfalls in the middle of the estuary. The concentration of slowly decaying detritus appears to be fairly uniform within the estuary and is related to the suspended mud concentration which is almost an order of magnitude higher than that found in the Severn Estuary at Hinkley Point.

The numerical studies have shown that the dissolved oxygen levels in the Parrett Estuary would naturally fall to about 60% of the saturation value during spring tides in the absence of any artificial BOD loads into the estuary.

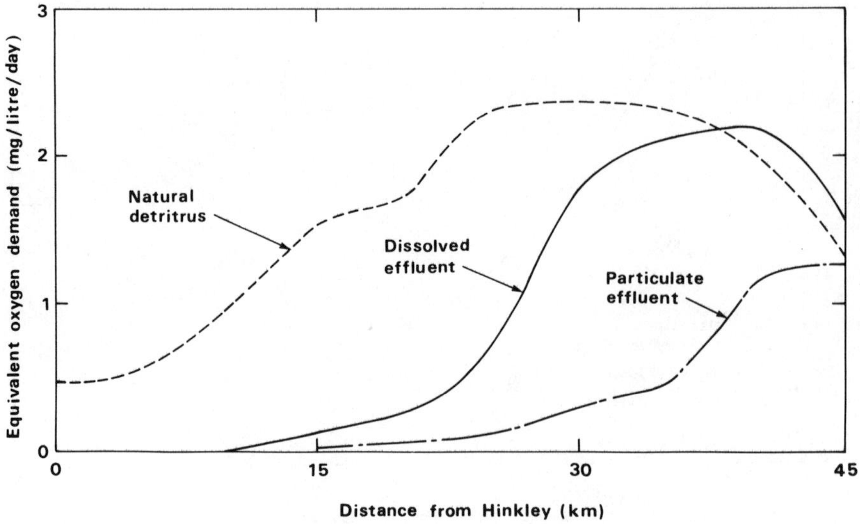


Fig. 7 Computed longitudinal variation of oxygen demand

The author believes that it is now a practical proposition to include the transport and oxidation of natural detritus and particulate effluent material in mathematical models of the oxygen balance of estuaries in one, two or even three-dimensions.

## ACKNOWLEDGEMENTS

The research described in this paper was funded by the U.K. Department of the Environment under contract PECD 7/7/049. The work was carried out by the author at Hydraulics Research Ltd., Wallingford, U.K., under the direction of Mr. N.V.M. Odd, Manager of the Tidal Computations Group in the Tidal Engineering Department headed by Mr. M.F.C. Thorn.

## REFERENCES

Barrett M.J., Gilligan R.M. and Mollowney B.M. 1972. Degradation of sludge. Out of Sight, Out of Mind. 1972 Report on the disposal of sludge in Liverpool Bay, H.M.S.O., 1972.

Maskell J.M., 1984 A one-dimensional model of water quality in tidal channel networks. Hydraulics Research Report 1984 IT 207.

Radford P.J., 1981. Modelling the impact of a tidal power scheme upon the Severn Estuary ecosystem. 1981 International symposium on Energy and Ecological Modelling, Louisville, Kentucky, 1981.

Welsh Water Authority, 1981.The Usk Estuary; Volume 1, an assessment of sewage disposal facilities. Tidal Waters Report No. 81/7.

# HYDRAULIC ASPECTS OF MODELLING WATER QUALITY IN DEEP ESTUARIES AND ENCLOSED BAYS

Nicholas V. M. Odd and James G. Rodger

Hydraulics Research Station, Wallingford OC10 8BA
United Kingdom

## ABSTRACT

The physical causing and modifying unsteady gravitational circulation in deep estuaries and enclosed bays is described with examples. The manner in which the gravitational circulation controls water quality in a deep estuary is illustrated with special reference to the dissolved oxygen distribution in the Tyne Estuary in 1969.

The failings of one-dimensional and depth-averaged models are assessed and the problems and performance of various types of multi-dimensional models are discussed.

## INTRODUCTION

The Mersey, Tyne and Tees, which are three of the most polluted tidal rivers in the U.K. (National Water Council, 1980), can be categorised as deep estuaries. The Santos Estuary in Brazil, The Rotterdam Water Way in Holland the Tejo Estuary in Portugal are also deep estuaries with significant pollution problems. Liverpool Bay in the U.K. and Tolo Harbour in Hong Kong are example of deep bays with brackish water and potential pollution problems.

The depth of an estuary is measured in absolute terms and relative to the tidal range. For practical purposes, one can assume that a deep estuary is one with mean depths in excess of about ten metres and a tidal range to mean depth ratio of less than about 0.5. The tide usually propagates through the deep regions of such estuaries without significant distortion due to shallow water effects.

## GRAVITATIONAL CIRCULATION

One of the most important aspects of the hydraulics of deep estuaries is the longitudinal gravitational circulation that is driven by the longitudinal density gradients within the estuary.

The magnitude of the net longitudinal pressure gradient, $d\rho/dx$, at a depth, z, which causes the gravitational circulation is a function of the slope of the mean tide level, $d\eta_0/dx$, and the vertical variation in the tide averaged longitudinal density gradient, $d\rho/dx$, as follows:

$$\frac{d\rho}{dx_z} = g\bar{\rho}_s\frac{d\eta_0}{dx} + g\int_z^{\eta}\frac{d\bar{\rho}}{dx}.dz \qquad 1$$

where $\rho_s$ is the density of surface water (Kg/m$^3$), and g is the acceleration of gravity (9.81 m/s$^2$)

The strength of the gravitational circulation varies directly with the magnitude of the product of the depth and the longitudinal density gradient. It is reduced by vertical mixing, which is usually heavily damped in stratified flows, and by energy dissipation at the bed, which is increased by the occurrence of high tidal velocities in the lower layers. The presence of a longitudinal density gradient within an estuary causes the mean tide levels to rise in a landward direction. The net landward pressure gradient and the net landward residual flow disappear at a 'null point' in the estuary where the two terms on the right hand side of equation (1) cancel each other out. The pattern of the gravitational circulation will vary according to the degree of stratification, but it is not dependent on the existence of vertical density stratification. Many deep estuaries with weak or negligible vertical stratification have strong gravitational circulations. The longitudinal density gradients tend to distort the shape of the velocity profile on the flood and ebb phases of the tide and thereby induce a net landward longitudinal amovement of water in the bed layers seaward of the flow of water in the surface layers of the estuary giving rise to a two-layer circulation.

### Mersey Estuary, U.K.

The vertical distribution of the tide-averaged residual velocities in the deep part of the Mersey Estuary, shown in Fig. 1, is typical of many deep partially-stratified estuaries. The landward velocities are almost uniform over the depth of the lower layer but the seaward velocities in the upper layer increase linearly towards the surface. The level at which the residual currents reverse direction varies with the position in the estuary. If the tidal range is significant compared to the mean depth, there may be a three layer structure with a small net landward movement of water in the layers above mean tide level.

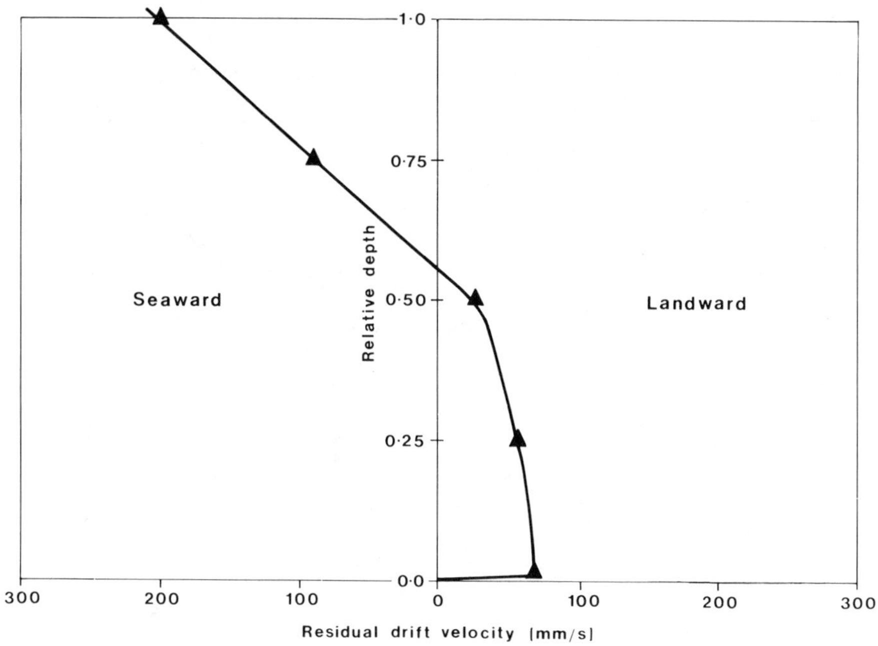


Fig. 1 Gravitational circulation in the Mersey estuary, U.K.

Tolo Harbour, Hong Kong

Numerically simulated longitudinal and vertical distributions of the tide averaged values of salinity and the gravitational circulation in Tolo Harbour, a long deep enclosed bay in Hong Kong, eight days after a flash flood is shown in Fig. 2.

The discharges involved in a gravitational circulation are often large compared to the fluvial discharge and the volumes of water moved within a tidal cycle may also be significant compared to the tidal volume.

In the case of the Mersey Estuary the landward discharge in the lower layer in the narrows is about 1,000 cumecs when the river discharge is 50 cumecs.

The strength of the gravitational circulation builds up fairly rapidly during the evacuation of a fluvial flood and then decays relatively slowly. The decaying circulation generated by one fluvial flood is often reinforced by the next flood as shown in Fig. 3. As a result, a fairly steady gravitational circulation may be maintained in an estuary or a bay fed by intermittent flashy fluvial floods. In some tropical situations the gravitational circulation generated by the fluvial floods in the wet season may persist right through the dry season.

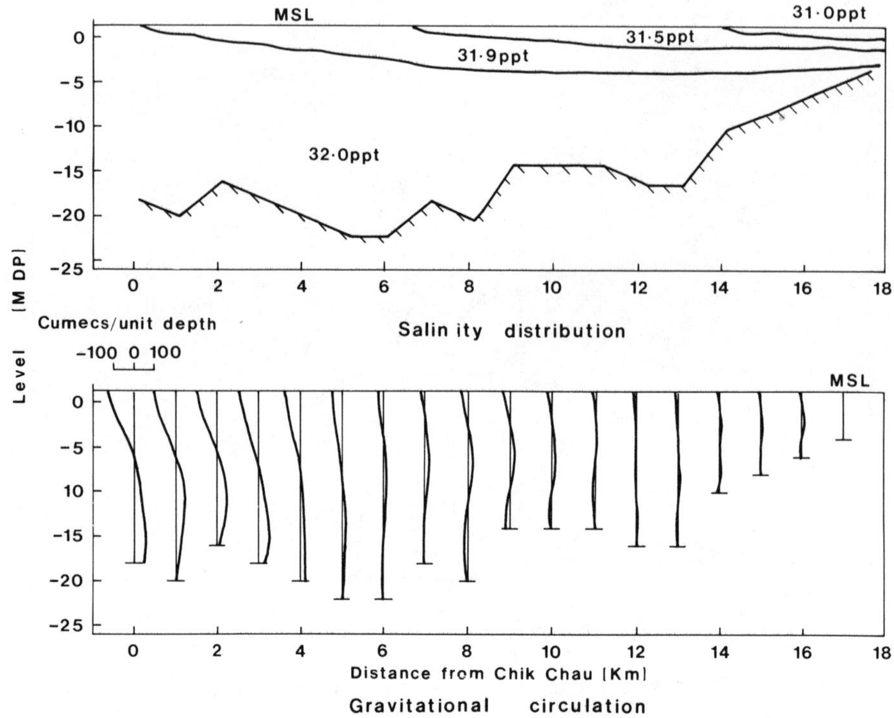


Fig. 2 Simulation of the vertical structure in Tolo Harbour

The presence of a strong two-layer gravitational circulation plays an important role in the transport and the vertical and horizontal distribution of suspended and dissolved matter in tidal flows. It, therefore, has a strong influence on the water quality (and aquatic life) in deep estuaries.

## Tyne Estuary, UK

Consider the tidally-averaged distribution of dissolved oxygen in the estuary of the River Tyne in 1969 (James, 1972) shown in Fig. 4. At that time the water temperature was relatively high at 15.8$^0$C and sewage discharges exerted their main biological oxygen demand in the landward and centre regions of the estuary. The fairly weak gravitational circulation, which was hindcast by numerical model techniques, introduces about 72 ($m^3$/s) of clean seawater into the lower layers of the estuary near the mouth. The oxygen content of the water in the lower layer, already limited by the water temperature, decreases in a landward direction as the oxygen demand in the water and from the bed takes effect to reach a minimum in the lower layers near the landward limit of the deep channel and the gravita-

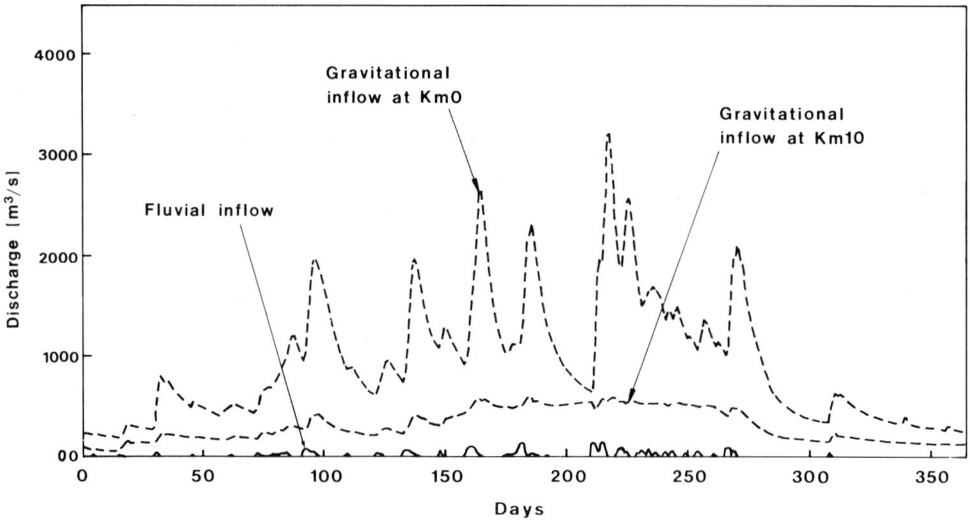


Fig. 3 Simulation of seasonal variation of gravitational flows in Tolo Harbour

tional circulation. The saline stratification damps vertical mixing, which inhibits the downward movement of dissolved oxygen from the surface layers. The oxygen deficient water in the lower layers upwells and mixes with the outgoing surface flow.

## Suspended matter

The gravitational circulation gives rise to different distributions for dissolved and suspended particulate matter. The behaviour of suspended matter is modified by the settling, deposition and re-erosion of the particles. Suspended particles leaving the estuary tend to settle into the lower layer at low water slack, and the gravitational circulation tends to return and trap a large proportion of suspended matter in an estuary (Price and Kendrick, 1963).

## One-dimensional or depth-averaged numerical models

The gravitational circulation can only be simulated in a numerical model that takes into account the vertical structure of the flow. One-dimensional area-averaged or two-dimensional depth-averaged numerical models are inadequate for simulating or predicting conditions in such estuaries, (Brady, 1980) because they provide no information about the vertical structure of the velocity, density and concentration profiles.

The results from one-dimensional models are also very sensitive to the prescribed value of the effective coefficient of longitudinal

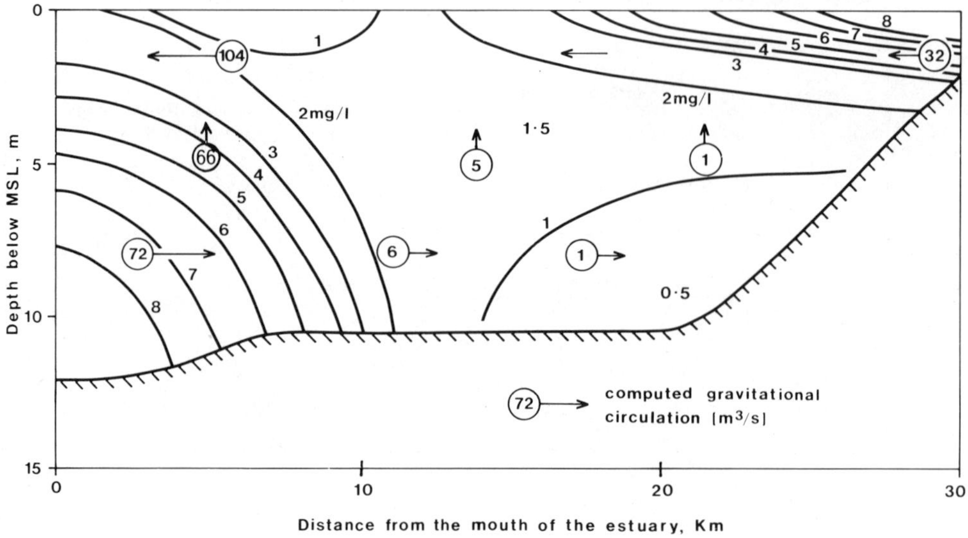


Fig. 4 Conditions in the Tyne estuary, August 1969.

dispersion, Dx, which varies with time and distance along an estuary. Its value varies in a highly non-linear fashion during the tidal cycle and with the fluvial discharge and tidal range. The manner in which the fluvial flows mix with seawater is unique to each estuary. There is no reliable method of predicting the value of Dx without recourse to detailed salinity observations for a wide range of fluvial and tidal conditions. Even this empirical method fails if engineering works change the geometry of the estuary or if predictions are required outside the range of observed conditions.

The common practice of increasing the magnitude of the coefficient for effective longitudinal dispersion in a one-dimensional model to account for the effect of gravitational circulation is unsatisfactory because it fails to reproduce the correct flushing mechanisms by which clean seawater is continually being introduced into the lower layers of the estuary.

## Multi-dimensional numerical models

The need for a two-dimensional in-depth (X-Z-T) representation of the Tees Estuary was recognised by ICI more than a decade ago when they developed a two-dimensional model of the transport of pollutants in the Tees (Hobbs and Fawcett, 1972) in which the tidal and gravitational flows were prescribed from observations. However, the results of this advanced model were very sensitive to the values of the prescribed co-efficients of vertical mixing and the prescribed gravitational circulation, which was difficult to quantify from observations.

Heaps and Jones (1977) successfully calculated the gravitational circulation in the Irish Sea using a three-dimensional model in which the horizontal density gradients were prescribed from salinity observations. There have been several attempts at using two-dimensional in-depth (X-Z-T) models to simulate tidal flows, gravitational circulation and saline intrusion in deep estuaries. However, many of these studies were hampered by the inability to prescribe the rate of vertical mixing in stratified tidal flows (Bowden and Hamilton, 1975). In recent years, the authors (Odd and Rodger, 1978) have devised a set of analytical mixing length functions derived from the theories of Prandl, Rosby and Montgomery, and Ellison as a means of calculating the internal shear stresses and vertical flux of solutes in gradually-varying turbulent stratified flows. These functions have been successfully implemented in a two-dimensional width-averaged model (Rodger, 1980) to simulate mud transport in the Brisbane River estuary (Odd and Baxter, 1980). This generalised model is capable of predicting conditions in any estuary which is narrow with respect to its length. An example of the use of this model to hindcast the gravitational circulation generated in the Tyne estuary is shown in Fig. 4.

In the authors' opinion a knowledge of the physical processes causing the transport and dispersion of both dissolved and suspended matter in stratified flows has now reached a stage when it is practical to use fully interactive two-dimensional (in depth) or two-dimensional (in plan) two layer model to simulate tidal flows, saline intrusion, mud transport, oxygen balance, the movement of heavy metals and the ecosystem in deep estuaries. However, since multi-dimensional inter-tidal models are relatively expensive to run, the authors have developed a quasi three-dimensional model, with large elements, which can simulate the unsteady gravitational circulation in an estuary or a enclosed bay for periods of up to a year as shown in Fig. 3. The results of this model have been used to drive a water quality and ecosystem model of Tolo Harbour in Hong Kong.

ACKNOWLEDGEMENTS

This paper describes research carried out by the Hydraulics Research Station, Wallingford and is published with permission. The work of preparing the paper was funded by the Department of the Environment under Contract PECD/7/7/049. The tests using the X-Z-T model were funded by the Department of Transport under Contract DGR/465/35. The models of Tolo Harbour were developed by the Hydraulics Research Station in association with Binnie and Partners, and funded by the Government of Hong Kong.

## REFERENCES


Bowden, K.F. and Hamilton, P. 1975. Some experiments with a numerical model of circulation and mixing in a tidal estuary. Estuarine and Coastal Marine Science, 3, 281-301.

Brady, J.A. 1980, Operational use of estuary models by the Northumbrian Water Authority. Paper No. 4 Water Research Centre/Institute of Hydrology Colloquia on Residual flows in rivers and estuaries, 20 October, 1980, Henley, U.K.

Heaps, N.S. and Jones, J.E. 1977. Density currents in the Irish Sea. Geophysical Journal of the Royal Astronomical Society, 51, 393-429.

Hobbs, J.D. and Fawcett, A. 1972. Two-dimensional estuary models. Paper No. 12. Symposium on mathematical and hydraulic modelling of estuarine pollution, Stevenage 1972.

James, A. (Ed) 1972. Pollution of the River Tyne Estuary. Bulletin No. 42. Department of Civil Engineering, University of Newcastle-upon-Tyne, 1972.

National Water Council 1980. River Quality the 1980 survey and future outlook. National Water Council, London.

Odd, N.V.M. and Rodger, J.G. 1978. Vertical mixing in stratified tidal flows. Journal of the Hydraulics Division HY3, American Society of Civil Engineers, 78.

Odd, N.V.M. and Baxter, T. 1980. Port of Brisbane siltation study 17th Conference on coastal engineering, Sydney 1980.

Price, W.A. and Kendrick, M.P. 1963. Field and model investigation into the reasons of siltation in the Mersey Estuary. Proceedings of the Institute of Civil Engineers 24, 473-518.

Rodger, J.G. 1980. Simulation of stratified flows in estuaries. 2nd IAHR Symposium on stratified flows, Trondheim, 1980.

# FIELD MEASUREMENT OF THE BIOLOGICAL IMPACT OF POLLUTION

J. Widdows

Institute for Marine Environmental Research
Prospect Place, The Hoe, Plymouth, PL1 3DH, England

## ABSTRACT

Field measurement of the biological responses of marine animals to pollutants has an important role in environmental management. Chemical monitoring provides an assessment of the degree of environmental contamination, but only biological monitoring can measure the impact on the environment.

The conceptual framework for biological effects measurement in environmental monitoring is outlined. Biological effects monitoring may in principle be carried out at any level of biological organisation (ecosystem, community, population, organismic, cellular and subcellular) by measuring aspects of their structure and function. At present, however, there is insufficient understanding of pollution effects at these higher levels of organisation to provide techniques for the early detection of any gradual deterioration or improvement in the environment. The need for more sensitive measurements of adverse biological effects of environmental contamination is met by the structural and functional responses at the lower levels of organisation.

Marine bivalves are commonly employed as 'sentinel organisms' in chemical and biological effects monitoring programmes. The results from some recent field studies that have used sublethal biological stress indices to measure the impact of pollution, are discussed.

Finally, the role of biological effects measurements in predictive modelling of 'environmental quality' or 'assimilative capacity' of marine environments is discussed.

# THE ROLE OF BENTHIC ALGAE IN ESTUARINE POLLUTION ASSESSMENT

Martin Wilkinson and David A. Rendall

Department of Brewing and Biological Sciences
Heriot-Watt University
Chambers Street
Edinburgh EH1 1HX
United Kingdom

## INTRODUCTION

Benthic algae might be considered suitable for estuarine pollution assessment for two reasons. Firstly they can be more abundant in the upper reaches of some estuaries than are the more commonly used macrofauna (but not necessarily the meiofauna). Secondly they present a more economic approach being quickly and cheaply sampled compared wth meiofauna and macrofauna. Although they are believed by some to be a notorious example of taxonomic confusion this is largely a reflection of their lack of study. Relatively few species are involved and their identification can be learnt without too much difficulty.

Methods for using the algae are better developed in freshwater practice and some of the methods used there might be found to be suitable for use in estuarine waters. A good summary of these is given by Thorpe and Williams (1980). Few of the suggestions for using algae made up to the present time have been refined to workable protocols and so, rather than surveying current practice, this paper gives some of the possible lines for future investigations.

## NATURE OF THE ALGAE

Characterisitics of the benthic macroalgal vegetation of estuaries will determine its possible use in pollution assessment. A

Slime of bacteria, diatoms and blue-green algae

↓

Fast growing green algae, mainly simple filamentous forms

↓

Foliose green algae and filamentous brown algae

↓

More complex brown and red algae of intermediate size

↓

Climax vegetation of fucoids

Fig. 1 A generalized ecological succession of algae colonizing a British sheltered rocky shore.

summary of relevant features is given below. A more detailed account is available in Wilkinson (1980).

The algae being considered are those that form macroscopically visible growths on hard substrata. These are the red, green and brown seaweeds and microscopic members of the Xanthophyceae, blue-green algae and diatoms that form visible mats.

In estuaries there are two broad physiognomic types of benthic algal flora. The lower reaches tend to have rocky shores with an impoverished marine flora dominated by large seaweeds, such as fucoids, but with relatively few species of red algae. Such vegetation can be monitored in the conventional ways applied to coastal seaweeds such as bioaccumulation monitoring system (Nelson-Smith, 1979). The upper reaches, where pollution problems are greater, are characterised by macroscopic mats of green, xanthophycean and diatom microalgae of mat types, which is most apparent in highly stratified estuaries (Wilkinson et al, 1981). This paper will describe problems arising from this paucity of species.

While the fucoid-dominated flora of lower estuaries arises through the ecological succession typical of sheltered epilithic algal communities (Fig. 1) the upper estuarine mat-forming flora appears to represent the succession arrested at an early stage. Therefore detection of disturbance by presence of early seral stages (Lewis, 1977) is not possible in the upper estuarine flora.

## USES OF ALGAE IN MONITORING

It has been suggested that algae can be used for the direct measurement of pollutants through bioaccumulation; for classification of estuaries and of reaches within estuaries; and for the monitoring of ecological effects of discharges either qualitatively or quantitatively using indicator species or communities.

## BIOACCUMULATION MONITORING

Seaweeds have become well established for monitoring certain pollutants; hydrocarbons, heavy metals, pesticides and radionuclides; on the open coast by virtue of their ability to accumulate measurable levels of these substances in their tissues from trace levels in the water column. The subject is reviewed by Phillips (1980). Certain slow growing algae, particularly _Fucus_ spp. and _Ascophyllum_ have been regarded as useful for time-integrating dissolved as opposed to particulate levels, particularly of heavy metals. However, a recent paper (Luoma et al, 1982) casts doubt on the sensitivity specifically to dissolved levels. Firstly the species with which it has been successfully applied do not penetrate far enough upstream. Secondly the characteristic upstream species are of a fundamentally different life-form to the successful monitors. They are fast growing, short-lived species of simpler physiology and morphology. Many of them are not thought to live as long as the time-scale of metal accumulation by the fucoids while their different nature may mean that the characteristics of the accumulation process are different from those in fucoids. They are more polymorphic making identification of specimens for monitoring by non biologically trained personnel more difficult than for fucoids More particularly this raises the question of the existence of different physiological races, which may have different accumulation characteristics, within the one species. They occur in mats which are often mixed species stands meaning that accumulation characteristics need investigating for an assemblage of species. The mats are in much more intimate contact with the sediment than are fucoid communities so that there is the possibility that upper estuarine algae might respond even more to adsorbed as well as dissolved metal levels. It is clear that much work needs to be done on the extension of this monitoring to upper estuarine algae, but it is encouraging that preliminary work by the North-West Water Authority (R.G. Chambers - personal communication) seems to indicate that _Enteromorpha_ spp. may prove suitable.

## CLASSIFICATION OF ESTUARIES USING ALGAE

Scandinavian workers (e.g. den Hartog, 1967) have produced lists of species of algae representative of different estuary zones within the Venice system for the classification of brackish-water. This may

Table 1 Common species of intertidal benthic algae which are widely distributed between British estuaries in the mid and upper reaches.

| | |
|---|---|
| CHLOROPHYCEAE | PHAEOPHYCEAE |
| Blidingia marginata | Ectocarpus siliculosus |
| B. minima | Fucus ceranoides |
| Cappsosiphon fulvescens | F. spiralis |
| Cladophora glomerata | F. vesiculosus |
| Enteromorpha intestinalis | Pilayella littoralis |
| E. prolifera | |
| E. torta | RHODOPHYCEAE |
| Monostroma oxyspermum | Audouinella purpurea |
| Prasiola stipitata | Bangia fuscopurpurea |
| P. crispa | *Porphyridium sp. |
| Percursaria percursa | BACILLARIOPHYCEAE |
| Rhizocolnium riparium | Melosira spp. |
| Rosenvingiella polyrhiza | M. nummuloides |
| Ulothrix flacca | "Schizonema" |
| U. speciosa | |
| U. subflaccida | MYXOPHYCEAE |
| Urospora penicilliformis | Gloeocapsa spp. |
| | Lyngbya spp. |
| XANTHOPHYCEAE | Microcoleus spp. |
| Vaucheria compacta | Oscillatoria spp. |
| Vaucheria spp. | Nostoc spp. |
| | Phormidium spp. |
| EUGLENNOPHYCEAE | Rivularia spp. |
| *Euglena obtusa | Aphanocapsa spp. |
| | Anabaena spp. |

*Two species of unicellular algae have been included for completeness because they form distinctive coloured patches on sediments.

be applicable in stable salinity situations such as the Baltic Sea and in large continental estuaries such as the Rhine system, but it does not seem so applicable in the United Kingdom with smaller estuaries whose salinity may vary over a wider range at any one point. It has been applied to one large UK estuary, the Thames (Tittley and Price, 1977) but that required modification of the system by redefinition of the salinity limits of the classes. Our experience at Heriot-Watt University in a survey, over 10 years, of algae in 108 estuaries around Britain is that upper estuarine algae are widely tolerant of environmental conditions. There is a set of relatively few common species (Table 1) which are very widespread both within and between estuaries, making zone identification within the estuaries difficult, as well as making it difficult to distinguish different classes of estuaries on the basis of their algal flora. At an

Table 2 Classification of estuaries into 3 types on the basis of the algal vegetation of the mid and upper reaches (according to Wilkinson et al, 1976)

| | |
|---|---|
| Type A | Mainly Chlorophyta and *Vaucheria* spp. |
| Type B | Upper part of eulittoral zone occupied mainly by Chlorophyta and *Vaucheria* spp. but lower dominated by *Melosira nummuloides*, a centric diatom forming filaments aggregated together in dense brown mats. |
| Type C | Chlorophyta, *Vaucheria* spp. and very conspicuous blue-green algae forming gelatinous masses, e.g. *Rivularia* and *Nostoc*. |

early stage of that survey when only 15 estuaries had been surveyed (Wilkinson et al, 1976) it appeared that the estuaries could be divided into 3 floral types (Table 2). Type B was held to be indicative of high organic pollution. With 108 estuaries surveyed it is now realised that types A and B intergrade. Most estuaries are green dominated but in such estuaries *Melosira nummuloides* can be extremely abundant when organic pollution is severe. Most estuaries appear to possess the basic pool of common species (Table 1) but different ones are selected to be abundant in different estuaries. There is therefore a need to consider abundance as well as species presence in classifying estuaries.

Type C (Table 2) remains distinct. It is important to realise that this is characterised by certain blue-green algae only. Almost all estuaries surveyed seem to possess, to a greater or lesser degree, mats of filamentous blue-green algae such as *Phormidium* spp. where filaments can coalesce to produce large expanses of flat strata. Only type C (Table 2) possesses conspicuous nodule-like colonies of blue-green algae such as are formed by the genera *Nostoc* and *Rivularia*. All estuaries of the 108 studied, with this type of flora are relatively unpolluted estuaries with a peaty freshwater inflow. Cotton (1912) found such species abundant in such estuaries in western Ireland. Whether it is the unpolluted nature of the estuaries or the oligotrophic acidic input which is responsible for this species pattern is not known. In the highly complex environment of estuaries the explanation might be found in yet other features such as the salinity pattern. More investigation is needed.

## MONITORING OF ECOLOGICAL EFFECTS USING ALGAE.

Ecological effects monitoring can be carried out by studying the whole affected community or by using particular species regarded as indicators of particular conditions.

Pollution indicator species are virtually impossible to find among the estuarine benthic algae. As explained above there is a pool of relatively few species which are of very wide tolerance. Any use of single species as indicators is therefore doubtful and in any event would have to be based on the species being abundant rather than merely present. The centric diatom *Melosira nummuloides*, so abundant in organically-polluted brackish waters, might have been thought of as an indicator by its abundance. There is however the complicating factor of the rigorous and variable physical environment of estuaries. Wulff and McIntire (1972) showed experimentally that large variations in its abundance could be brought about in a laboratory stream by variable as opposed to constant, levels of salinity. The results of Rendall (1982) suggested that it might be as much adapted to a rapidly changing environment as to a polluted one.

An alternative candidate as a indicator species, by its abundance, is the filamentous green alga *Rosenvingiella polyrhiza*. This is the only species to be successful in the parts of the Garnock estuary most badly affected by low pH and metal enrichment (Wilkinson et al, 1980). Its unique dominance has also been recorded by us at a similarly polluted site on the Billingham Beck, an estuarine tributary of the River Tees. It is of little use to indicate the existence of the unfavourable environment that is so extreme in these examples (e.g. pH 2.0) that biological indication is unnecessary. In indicating the magnitude of the ecological effect of the discharges it is the absence of all the other species as much as the abundance of *Rosenvingiella* that is important.

The purpose of presenting the two examples above is to show that the use of single species in estuarine pollution assessment is fraught with difficulties. This is intensified by the possession by some estuarine algae of bimodal distributions, with abundances under opposite extremes of environmental conditions and little occurrence in intermediate conditions (Wilkinson, 1980). An exaple is the blue-green alga *Phormidium corium* reported as abundant under the cleanest and most polluted conditions but not so in waters of intermediate quality. For all the reasons above it seems that communities rather than single species of algae would be of use in estuarine pollution assessment.

Table 3 Totals of multicellular algal species found on 35 intertidal rocky shores sampled during British Phycological Society field meetings, 1969-1978, with comparable sampling effort and taxonomic expertise. (Shores in Co. Wexford, Eire (5 shores); Lewis and Harris (5); Galloway (6); Glamorgan (4); Orkney (4); Grampian (5); Berwick and East Lothian (6)).

| | Rhodophyta | Chlorophyta | Phaeophyta | Total |
|---|---|---|---|---|
| Maximum number of species | 59 | 24 | 45 | 110 |
| Minimum number of species | 31 | 10 | 17 | 68 |
| Mean number of species | 44 | 16 | 30 | 90 |
| Standard deviation | 6.9 | 3.8 | 7.3 | 10.4 |

The use of communities could be based qualitatively on the list of species present or quantitatively on numbers of species and/or their abundances.

The simplest measure that could be used is species richness, the total number of species present. This is likely to be of little value in upper estuaries where few species are present but in the fucoid dominated rocky shores of the lower reaches of large estuaries it might be so. Table 3 shows the totals of intertidal algae present on 35 different open coast shores in Britain. These have been deliberately chosen to represent sites comparable in terms of physical area, habitat diversty, taxonomic expertise in the work up, and sampling effort. With these factors controlled there is a relatively narrow-range of variation in species total. It is suggested that a species total much lower than this range would represent a disturbed habitat.

This is a suggestion made in the hope of promoting further investigation. It is not, at this stage, possible to specify by how much a species total should be reduced to indicate an adverse effect. Our preliminary assessment, based on too small a sample of sewage-polluted shores for inclusion in Table 3, suggests that badly affected shores may have a species reduction of the order of 50%. Setting of more precise statistical limits (e.g. 95% confidence limits) in order to detect more subtle effects is probably unrealistic and impossible in the present state of algal knowledge. This applies to

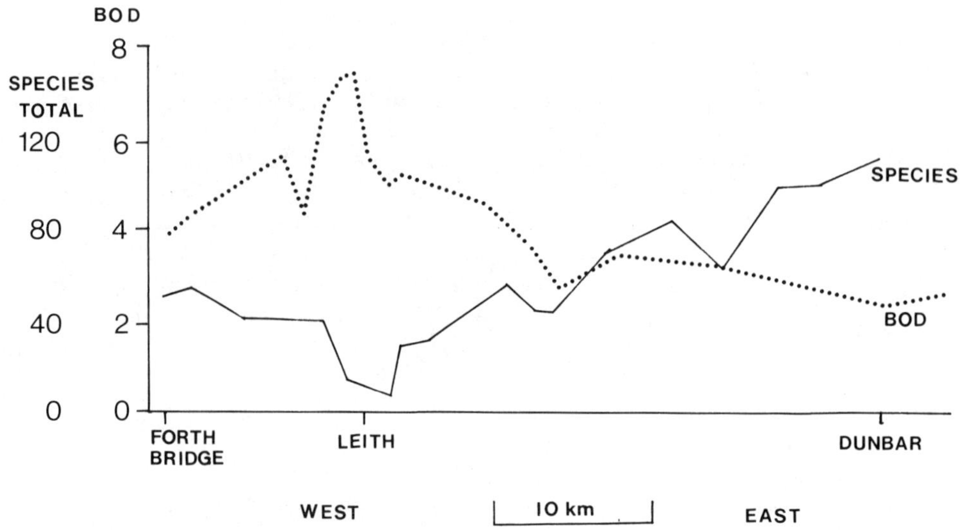


Fig. 2 The distribution of algal species total along the south bank of the Firth of Forth, with biochemical oxygen demand (BOD) in mg $l^{-1}$ as indicative of water quality, plotted for comparison along the same horizontal scale. (BOD readings are means of 12 values at each site at high tide 1977-78). The figure refers to the lower part of the Forth estuary where salinity is not a complicating factor, being generally above 30‰.

the open coast but could apply to the lower most reaches of estuaries. Figure 2 presents an example from the Firth of Forth. Note that localised areas of species depletion on the north and south banks correspond to areas of adverse water quality.

In addition to species total, abundance might be taken into account. Patrick (1964) used these two features to analyse the structure of diatom communities on artificial collectors from polluted and unpolluted environments. She plotted the distribution of numbers of species having a particular level of abundance against the levels of abundance based on geometric classes. The communities always gave the general shape of curve shown in Figure 3. However the mode was consistently depressed and the length extended under polluted conditions because the community became characterised by fewer more tolerant species present in greater abundance. Although Patrick's (1964) work is often associated with freshwater, she showed that this idea applied to brackish-water diatom communities and commented that because there was a large pool of microscopic pennate diatoms in estuaries this approach would be suitable in estuaries. This is worthy of further investigation although the difficulty of identification of

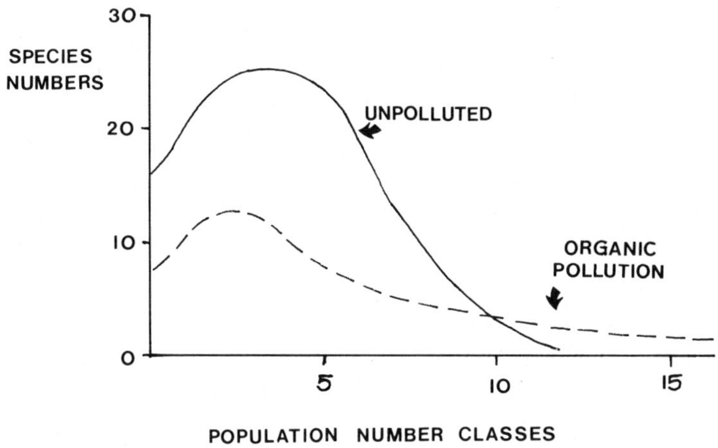


Fig. 3 Diatom community structure acccording to Patrick (1974). Geometric classes are based on successive doublings of population numbers, class 1 = 1-2 individuals class 2 = 3-4 individuals, class 3 = 5-8 individuals, class 4 = 9-16 individuals etc.).

pennate diatoms might limit its application. This could perhaps be overcome by using indices of diversity and basing these on the Sequential Comparison Index of Cairns et al (1968). Instead of using only numerical estimates of community structure some account could be taken of which actual species contributed to this structure i.e. species could be ranked according to pollution tolerance.

This has been done successfully in freshwater in, for example, the saprobic system of Fjerdingstad (1964) where different zones in recovery of a river from organic pollution are characterised by different algal and microbial communities. For each zone there are alternatives depending on the local natural water conditions e.g. hard or soft water. Such a system could be inapplicable to most UK estuaries because of the universality of species already described. Attempts have been made to construct saprobic classifications of species for larger bodies of brackish water, where species numbers are greater, such as the Baltic Sea (e.g. Hayren, 1923, Grenager 1957). A problem in this is that different workers have contradictory ratings for important species. This probably reflects our lack of knowledge of the other natural factors influencing the species distributions in estuaries.

Lindgren (reported in Wallentinus (1979)) has taken this a stage further by incorporating abundance. He has proposed an index of eutrophication which can be calculated for a brackish water macro- algal community and has shown that it can be related to water quality in an area of the Baltic Sea. This index is based on three simple categories of abundance and a classification of the algae into seven categories based on their reaction to eutrophication (3 positively influenced, one neutral, 3 negatively influenced). The product of the abundance rating and the reaction category for each species in the community is averaged to give the eutrophication status of the community. The application of such a system to UK estuaries, where abundance is as important as the list of species present, is worthy of investigation. Both ranking-type systems devised relate mainly to sewage pollution and application to industrial pollution would need to be investigated.

In this short paper it has not been possible to review all likely approaches. It is interesting that some of the approaches resemble very closely those of zoologists. For example Gray (1979) has considered, in relation to animal populations, the ideas both of short-lived opportunistic species and of departures from the log-normal distribution. The particular approaches described in this paper have been chosen as being based largely on field populations or being related to methods in use with animals and/or in freshwater. Others remain such as the testing of growth potential of waters for macro- and microalgae under standard laboratory conditions and the growth of cultures and of transplanted alage in the field. The lack of definitely established methods being described in this paper is a reflection of the lack of work in this field rather than of any inability of algae to be used.

## REFERENCES


Cairns, J., Albaugh, D.W.,Busey, F. and Chanay, M.D. 1968. The Sequential Comparison Index-a simplified method for non-biologists to estimate relative differences in biological diversity in stream pollution studies. _Journal of the Water Pollution Control Federation_ _40_, 1607-1613.

Cotton, A.D. 1912.Marine Algae. In _A Biological Survey of Clare Island in the County of Mayo, Ireland, and of the Adjoining District_ (Praeger, R.L. (Ed)) _Proceedings of the Royal Irish Academy_ _31_, (15), 1-178.

Fjerdingstad, E. 1964.Pollution of streams estimated by benthal phytomicro-organisms. I. A saprobic system based on communities of organisms and ecological factors. _Internationale Revue der gesamten Hydrobiologie_ _49_, 63-131.

Gray, J.S. 1979.Pollution-induced changes in populations. _Philosophical Transactions of the Royal Society of London B286_, 545-561.

Grenager, B. 1957.Algological observations from the polluted area of the Oslo fjord. _Nytt. Magazin for Botanik_ _5_, 41-60.

den Hartog, C. 1967.Brackish water as an environment for algae. _Blumea_ _15_, 31-43.

Hayren, E. 1923.Fororenigen och strandvegetationen i Helsingfors Hamnomrade. _Svensk Botanisk Tidskrift_ _17_, 62-668.

Lewis, J.R. 1977.Rocky foreshores. In The Coastline (Barnes, R.S.K. (Ed)). John Wiley & Sons, London pp. 147-158.

Luoma, S.N., Bryan, G.W. and Langston, W.J. 1982.Scavenging of heavy metals from particulates by brown seaweed. Marine Pollution Bulletin 13, 394-396.

Nelson-Smith, A. 1979.Monitoring the effect of oil pollution on rocky seashores. In Monitoring the Marine Environment (Nichols, D. (Ed)). Institute of Biology, London, pp. 25-53.

Patrick, R. 1964.A discussion of natural and abnormal diatom communities. In Algae and Man. (Jackson, D.F. (Ed)). Plenum Press, New York. pp. 185-204.

Phillips, D.J.H. 1980.Quantitative Aquatic Biological Indicators. Applied Science Publishers, Barking, England.

Rendall, D.A. 1982.Factors affecting the distribution of Melosira nummuloides in the polluted Clyde estuary. Ph.D. Thesis. Heriot-Watt University, Edinburgh, Scotland.

Thorpe, M.W. and Williams, I.L. 1980.A guide to the use of algae for the biological surveillance or rivers. Technical Memorandum No. 20, Department of the Environment, Water Data Unit, Reading, England.

Tittley, I. and Price, J.H. 1977.The marine algae of the tidal Thames. London Naturalist 56, 10-17.

Wallentinus, I. 1979.Environmental influences on benthic macrovegetation in the Trosa-Asko area, northern Baltic proper. II. The ecology of macroalgae and submersed phanerogams. Contributions from the Asko Laboratory No. 25, 1-210.

Wilkinson, M. 1980.Estuarine benthic algae and their environment - a review. In The Shore Environment Vol. 2. (Price, J.H., Irvine, D.E.G., and Farnham W.F. (Eds)). Academic Press, London. pp. 425-486.

Wilkinson, M., Henderson, A.R. and Wilkinson, C.1976. Distribution of attached algae in estuaries. Marine Pollution Bulletin 7, 183-184.

Wilkinson, M., Macleod, L.and Fuller, I. 1980. A first account of estuarine algae under combined conditions of very low pH and metal enrichment. Botanica marina 23, 457-477.

Wilkinson, M., Fuller, I.,Penny, J.W., Scanlan, C.W., May, R.L. and Roe, F. 1981. Vertical zonation of intertidal algae in some small stratified estuaries. Phycologia 20, 116-117.

Wulff, B.L. and McIntire, C.D. 1972.Laboratory studies of assemblages of attached estuarine diatoms. Limnology and Oceanography 17, 200-214.

# OLIGOCHAETES AS INDICATORS OF ESTUARINE POLLUTION

B.E. Barnett

Anglian Water Authority, Bridge Street, Brigg DN20 8NQ
England

ABSTRACT

The potential role of oligochaetes as pollution indicators in saline waters is considered in the light of recent literature describing oligochaete distributions in relation to pollution. The principle estuarine oligochaetes are identified and each is evaluated as possible pollution indicator species. Enchytraeid oligochaetes pose particular taxonomic problems and are therefore of limited value as pollution indicators at the present time. *Paranais litoralis* is the most commonly encountered estuarine representative of the family Naididae, but populations show very wide natural fluctuations and it is therefore difficult to relate its distribution and aundance to the effects of pollution. A second Naidid, *Amphichaeta sannio* has currently been recorded in only two British estuaries although it is probably quite widespread in estuarine habitats. Until more is known of the ecology of *A. sannio* the species can not be considered useful as a pollution indicator.

Three spcies of Tubificid; *Tubifex costatus*, *Tubificoides benedeni* and *T. pseudogaster* are widely reported from estuarine localities. *T. pseudogaster* does not appear to be very widespread and insufficient information exists to ascertain the value of this species as a pollution indicator. *T. costatus* has been reported as being associated with pollution in the less saline reaches of several estuaries. It may therefore be a potentially useful indicator species although its success in the areas examined may result more from its resistance to the natural rigours of the brackish water habitat than tolerance of pollution. *T. benedeni* is the dominant Tubificid in the more saline reaches of estuaries and in many coastal areas. The species has been widely reported from polluted situations and experimental evidence suggests that it is extremely pollution tolerant. It is therefore suggested that *Tubificoides benedeni* may be considered as a valid indicator species.

Interpretation of the distribution and abundance of indicator species is discussed with reference to dominance and diversity, and it is concluded that tubificid oligochaetes may have a useful role to play as indicators of pollution in saline waters.

# INVERTEBRATE PRODUCTION ECOLOGY IN RELATION TO ESTUARINE QUALITY MANAGEMENT

M. Elliott[1] and D.S. McLusky[2]

1. Forth River Purification Board
Estuary Laboratories, Port Edgar Marina
South Queensferry, Edinburgh EH30 9SQ
2. Department of Biology
University of Stirling, FK9 4LA
Scotland

## ABSTRACT

The use of the production ecology of estuarine macrobenthic invertebrates is proposed as a measure of detecting sub-lethal stress within those populations. It is suggested that the growth and amounts and rates of secondary production give a long-term integration of the environmental factors, both natural and polluting, affecting such populations. Furthermore, those parameters can be used to detect stress on an estuarine community at a level lower than that needed to change the community structure. A case-study of part of the industrialised Forth estuary, Scotland, is used to illustrate the variation within these parameters in a *Macoma balthica* (*L*) population attributable both to the usual environmental changes encountered over an intertidal area and to the influence of polluting sources.

The second part of the paper contrasts the use of production ecology in detecting stress on estuarine invertebrates, and therefore its use in estuarine management, with other indices of stress on and change in communities, populations and individuals which may be attributable to pollution.

## INTRODUCTION

Macrobenthic infauna have many attributes which make their study a cost-effective method of assessing estuarine quality by aiding in the identification of the effects of pollution. These attributes in-

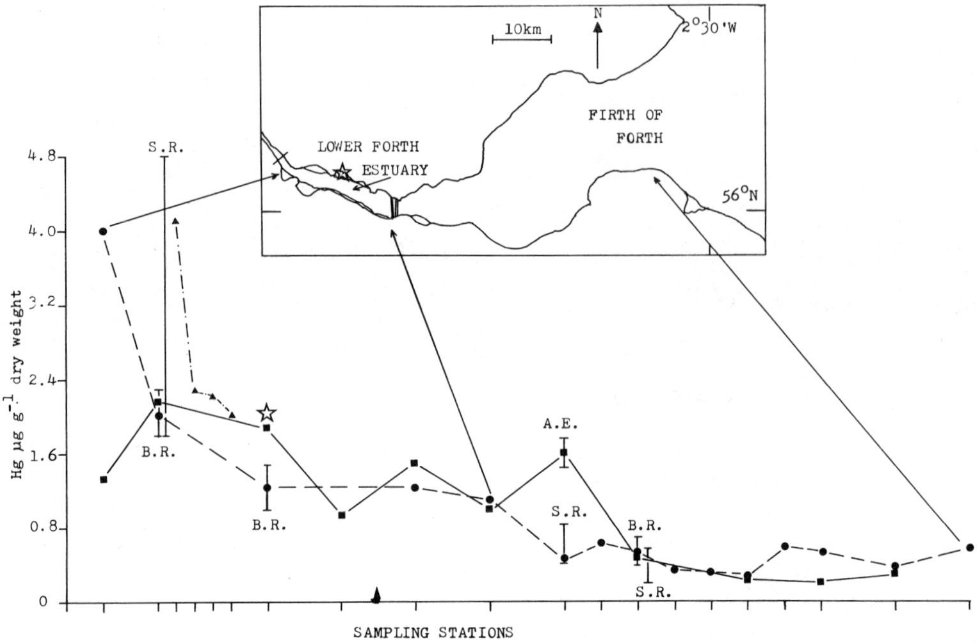


Fig. 1 Mercury Accumulation in the mussel, *Mytilus edulis* in the Estuary and Firth of Forth (1982 data). Symbols: ● = south shore; ■ = north shore; ▲ =mid channel; these symbols denote Hg levels in animals usually of length 40-45mm. S.R.= the range of values over the size range 30-70mm; B.R.=the range of 3 field replicates (i.e. animals of the same size); A.E.= analytical error. Subsamples from the homogenised tissue of approximately 15 animals were used for each analysis. The position of Torry Bay is shown by the star. (From Griffiths, Elliott and Dobson, in prep.)

clude their close association with the sediment, their relative inability to migrate, their often long-term integration of environmental effects and the relative ease and repeatability of sampling. An accepted environmental quality objective of river purification authorities is to ensure a suitable benthos for the support of indigenous fish populations; this has resulted in a large effort by those authorities to study stress within benthic populations. In this paper stress is defined as the cumulative quantifiable result of adverse environmental conditions or factors as an alteration in the state of an individual, population or community. Gray (1979) separates pollution effects into two types - stress and disturbance, the latter involving the physical destruction or removal of individuals from an area, whereas chemical stress "results in the productivity of an in-

Table 1. The major environmental variables at each station. (The salinity, %nitrogen, and % carbon data are the range of values encountered over the two year sampling period.)

| Station | Sediment | | | | % Nitrogen Range | % Carbon Range | Tidal Height above C.D. | Slope of Shore | Mean Immersion Period | Interstitial Salinity Range |
|---|---|---|---|---|---|---|---|---|---|---|
| | Median Particle Diameter (µm) | Sorting Coefficient (Ø units) | Skewness (Ø units) | % Silt and Clay | | | (m) | (°) | (hrs/tide) | (°/oo) |
| A1 | 41 | 1.49 | 0.228 | 79 | 0.131 - 0.434 | 3.03 - 7.19 | 3.36 | 8 | 6.18 | 28.10 - 33.75 |
| A2 | 123 | 1.40 | 0.297 | 32 | 0.066 - 0.166 | 1.26 - 3.89 | 3.97 | 8 | 5.09 | 26.52 - 33.48 |
| A3 | 55 | 1.93 | -0.157 | 55 | 0.050 - 0.151 | 1.06 - 4.56 | 4.20 | 6 | 4.22 | 28.00 - 33.15 |
| A4 | 216 | 1.25 | 0.083 | 14 | 0.086 - 0.160 | 1.62 - 3.62 | 4.50 | 11 | 3.44 | 27.00 - 33.38 |
| B1 | 39 | 1.44 | 0.527 | 84 | 0.165 - 0.433 | 3.38 - 6.74 | 2.60 | 5 | 8.08 | 25.62 - 33.80 |
| B2 | 44 | 1.40 | -0.024 | 76 | 0.141 - 0.436 | 3.55 - 7.95 | 2.97 | 2 | 7.07 | 27.32 - 33.95 |
| B3 | 176 | 1.57 | 0.368 | 30 | 0.122 - 0.447 | 2.07 - 7.34 | 3.09 | 4 | 6.77 | 25.00 - 32.75 |
| B4 | 210 | 0.88 | -0.510 | 3 | 0.013 - 0.069 | 0.22 - 1.26 | 3.50 | 9 | 5.71 | 22.62 - 33.63 |
| B5 | 180 | 1.06 | -0 132 | 3 | 0.042 - 0.106 | 0.71 - 2.27 | 4.13 | 12 | 4.45 | 27.37 - 34.53 |
| C1 | 57 | 1.75 | -0.202 | 54 | 0.200 - 0.380 | 4.70 - 6.61 | 1.27 | 6 | 10.64 | 28.22 - 33.07 |
| C2 | 178 | 1.66 | 0.352 | 21 | 0 024 - 0.379 | 0.58 - 6.36 | 2.15 | 7 | 10.38 | 22.47 - 34.00 |
| C3 | 234 | 0.75 | -0.072 | 4 | 0.024 - 0 121 | 0.31 - 1.96 | 2.71 | 18 | 7.67 | 28.22 - 33.15 |
| C4 | 35 | 1.07 | -0.194 | 1 | 0.029 - 0.199 | 0.30 - 3.24 | 3.96 | 10 | 4.83 | 22.32 - 34.65 |
| C5 | 388 | 1.53 | -0.347 | 2 | 0.031 - 0.126 | 0.48 - 2.84 | 4.56 | 18 | 2.84 | 24.55 - 35.28 |

dividual being reduced". The present authors suggest that disturbance may be regarded as community stress.

When using invertebrate populations to detect stress in the environment either of two sets of responses have been studied (i) structural responses, the measurement of biotic (taxonomic) or abiotic (non taxonomic) characteristics at a point in time, (ii) functional responses, the measurement of any rate process of the ecosystem (Matthews et al, 1982). Benthic studies applicable to pollution control and/or estuarine and coastal management have similarly included these two responses. Structural response studies include the numerous analyses of community structure and pollutant contaminaion in the biota; functional response studies include the rates of change of physiological parameters as an indication of stress (Bayne, 1978). Structural response studies are appropriate for detecting "catastrophic" changes in the environment, i.e. the change of a community due to the removal or replacement of species, normally found in a area, due to natural or anthropogenic causes (see Leppakoski, 1975; Pearson and Rosenberg, 1978; McLusky, 1982). Functional responses, however, are possibly more appropriate for detecting sublethal effects due to stress, i.e. for detecting pollution effects before the system is affected catastrophically.

In this paper the authors propose the use of production- ecology parameters of selected species in functional response studies as a means of assessing pollution-induced sublethal effects. Production ecology may be defined as that part of a species' ecology concerned with the analysis of parameters influencing production by the population, e.g. growth, somatic production, body condition changes, reproduction and productivity. Production is regarded here as the net amount of material fixed by the population, somatic production is the production attributable to growth. Productivity is used to describe the relationship between the population production and biomass often expressed as the P/B ratio. However, before such parameters can be used to compare geographical areas, it is necessary to assess the variability of those parameters within one area. The first part of this paper concerns a case-study of *Macoma baltica* (L) in the Forth Estuary; it both details that variability and gives some explanation for it while showing than an area may still be compared with other geographical areas. Since the International Biological Programme of the early 1970's, many productivity studies have resulted in large amounts of data on functional aspects of marine and estuarine ecosystems (e.g. Warwick and Price, 1975; Ansell et al, 1978; Robertson, 1979). That data allows comparisons to be made between different areas. The second part of the paper gives a review of biotic methods used to date in pollution assessment.

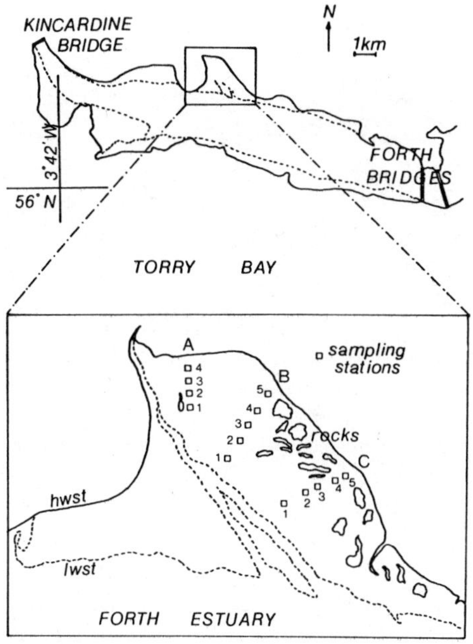


Fig. 2 Location of the lower Forth Estuary (top) and Torry Bay (bottom).

## A CASE-STUDY OF MACOMA IN THE FORTH ESTUARY

The growth, production and productivity of the infaunal selective-deposit and suspension feeding bivalve Macoma balthica (L) has been studied at Torry Bay, a 3.5 km$^2$ heterogeneous, intertidal mudflat on the northern shore of the Forth Estuary, eastern Scotland. McLusky and Elliott (1981) have described the principal faunistic components of Torry Bay; McLusky et al., (1976, 1978) give details of the intertidal fauna and industrial usage of the whole estuary, and McLusky (1982) detailed the community structural response to point-source, industrial pollution in one area of the estuary. As an indication of the polluted nature of the estuary, the levels of mercury accumulated in the mussel Mytilus edulis L. are given in Fig. 1, where it may be seen that despite the distance (over 6 km) from the major industrial discharges, mussels from Torry Bay still have high levels in relation to those close to the industrialised area.

In the study reported here a population of M. balthica was monitored over a 2 year period at 14 stations (Table 1, Fig. 2). The full details of the study, which was a development from a preliminary study reported by McLusky and Allan (1976), are given by Elliott (1979).

The growth of M. balthica at each station was assessed by growth - cessation ring analysis, size-frequency histogram interpretation

Table 2. The Bertalanffy growth equations and parameters for M. balthica at each station.

| Station | Equation | $H_1$ (mm) | $H_2$ (mm) | $H_4$ (mm) | $H_6$ (mm) | $H_8$ (mm) | $T_{95}$ (yrs) | $H_4$ Tdw (mg) | $H_4$ Fdw (mg) | $H_8$ Tdw (mg) | $H_8$ Fdw (mg) | Max. Size Sampled (mm) |
|---|---|---|---|---|---|---|---|---|---|---|---|---|
| A1 | $H_t = 16.002\,(1 - e^{-0.224(t - 0.271)})$ | 2.41 | 5.14 | 9.06 | 11.57 | 13.17 | 12.84 | 95.16 | 13.574 | 355.37 | 42.331 | 13.7 |
| A2 | $H_t = 10.450\,(1 - e^{-0.395(t - 0.103)})$ | 3.12 | 5.51 | 8.12 | 9.43 | 9.99 | 7.57 | 73.67 | 8.815 | 149.61 | 16.151 | 11.3 |
| A3 | $H_t = 10.134\,(1 - e^{-0.383(t - 0.290)})$ | 2.41 | 4.87 | 7.69 | 9.00 | 9.60 | 7.99 | 58.47 | 7.015 | 132.09 | 14.337 | 10.6 |
| A4 | $H_t = 10.670\,(1 - e^{-0.323(t + 0.120)})$ | 3.24 | 5.29 | 7.85 | 9.19 | 9.89 | 9.33 | 59.99 | 9.167 | 137.42 | 20.969 | 10.2 |
| B1 | $H_t = 16.970\,(1 - e^{-0.166(t + 0.134)})$ | 2.91 | 5.06 | 8.43 | 10.84 | 12.57 | 17.92 | 94.24 | 12.239 | 398.45 | 40.169 | 15.9 |
| B2 | $H_t = 14.020\,(1 - e^{-0.307(t - 0.509)})$ | 1.96 | 5.15 | 9.22 | 11.42 | 12.61 | 9.81 | 111.24 | 12.268 | 393.19 | 40.170 | 14.9 |
| B3 | $H_t = 12.981\,(1 - e^{-0.336(t - 0.228)})$ | 2.97 | 5.82 | 9.33 | 11.11 | 12.03 | 8.99 | 127.79 | 16.128 | 337.94 | 41.040 | 15.1 |
| B4 | $H_t = 12.038\,(1 - e^{-0.280(t - 0.320)})$ | 2.09 | 4.52 | 7.74 | 9.58 | 10.64 | 10.85 | 65.06 | 7.441 | 205.92 | 21.215 | 12.6 |
| B5 | $H_t = 12.270\,(1 - e^{-0.288(t - 0.410)})$ | 1.92 | 4.51 | 7.91 | 9.82 | 10.89 | 10.50 | 77.11 | 9.802 | 267.93 | 29.134 | 13.3 |
| C1 | $H_t = 15.038\,(1 - e^{-0.271(t - 0.278)})$ | 2.67 | 5.61 | 9.55 | 11.85 | 13.18 | 10.93 | 154.37 | 17.018 | 555.72 | 44.108 | 16.1 |
| C2 | $H_t = 12.097\,(1 - e^{-0.295(t - 0.263)})$ | 2.36 | 4.85 | 8.08 | 9.87 | 10.86 | 10.50 | 65.15 | 9.395 | 175.04 | 24.301 | 14.6 |
| C3 | $H_t = 13.508\,(1 - e^{-0.234(t - 0.329)})$ | 1.96 | 4.37 | 7.79 | 9.92 | 11.26 | 12.61 | 73.22 | 8.378 | 304.03 | 27.514 | 13.2 |
| C4 | $H_t = 11.626\,(1 - e^{-0.287(t - 0.083)})$ | 2.69 | 4.92 | 7.85 | 9.50 | 10.43 | 10.60 | 80.17 | 7.514 | 233.40 | 18.611 | 12.7 |
| C5 | $H_t = 9.220\,(1 - e^{-0.411(t - 0.148)})$ | 2.72 | 4.91 | 7.33 | 8.39 | 8.85 | 7.41 | 52.27 | 6.397 | 100.27 | 11.584 | 9.6 |

Where $H_1$, $H_2$, $H_4$, $H_6$ and $H_8$ = Heights at ages 1, 2, 4, 6 and 8 years respectively; $T_{95}$ = age at which 95% of maximum size is reached; $H_4$ Tdw, $H_4$ Fdw, $H_8$ Tdw, $H_8$ Fdw = total and flesh weight of 4 and 8 year animals, respectively.

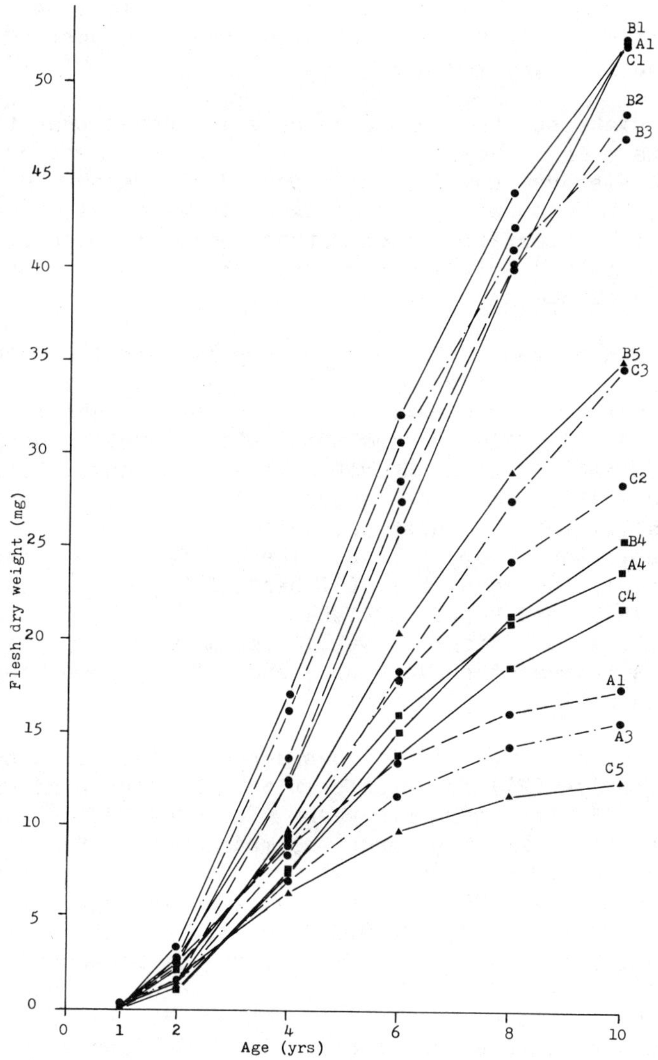


Fig. 3 Plots of the flesh-weight transformed Bertalanffy equations of M. balthica growth at each station.

and mark-and-recapture techniques. Using a two-step quantification of the resultant growth curves, a Bertalanffy equation (Beverton and Holt, 1957) and other growth parameters were then calculated (Table 2). The data show the large range over the Bay in the theoretical maximum shell height (H) and age ($T_{95}$) as 9 - 17 mm and 7 - 18 years respectively. The total and flesh dry weight (e.g. for four ($H_4$) and eight ($H_8$) year old animals) was then calculated for the size at those ages using winter flesh condition factors for each station. (Summer condition factors gave an even larger variation between the stations). The winter variation in eight-year old aimals was 11.6 to 44.1 mg flesh dry weight. The resultant weight-transformed Bertalanffy equations are given in Fig. 3.

Multivariate statistical techniques indicated that the growth parameters were influenced principally by the sediment physical structure parameters, the immersion period and hence the available food supply. Low-shore individuals were long-living, had lower values of K (the "coefficient of catabolism", the rate at which H was reached) and reached a larger size than those animals found on the upper regions of the shore.

The growth curves of M. balthica populations from other studies (Table 3, Fig. 4) can be grouped into two - those with a high growth rate, which reach a large maximum size but have a short life-span (populations 1 - 11, Fig. 4) and those with a smaller growth rate, which reach a smaller size but yet have a much longer life-span (populations 12 - 15). The first group include mainly temperate, intertidal or shallow water populations while the second is composed of arctic or sublittoral populations. These differences have been attributed mainly to temperature (Gilbert, 1973; Green, 1973) although Segestrale (1960) and McErlean (1964) consider that the variation reflects nutritive conditions as well. The mean growth pattern for Torry Bay was between the two groups and notably lower than for other comparable intertidal populations.

The annual somatic production and productivity (as production (P) / mean biomass (B)) for each station was calculated using method 2 of Crisp (1971) for each of 2 years (Table 4). The B and P varied over the area by at least an order of magnitude (0.820 - 9.095 g fdw $m^{-2}$ and -0.290 - 4.395 g fdw $m^{-2}$ $yr^{-1}$ for 1975-6 respectively, 0.481 - 5.692 and 0.0566 - 2.550 for 1976-7). The average values for the Bay were B 3.190, P 1.130, P/B 0.354 for 1975-6, and 1.880, 0.920, 0.489 for 1976-7 respectively. Some of the differences between 1975-6 and 1976-7 are due probably to the use of a longer sampling interval (monthly and quarterly, respectively). An estimate of this difference was calculated for one station (B3); the P, B and P/B on a monthly and quarterly basis for 1976-7 were 4.462, 1.839 and 0.412, and 3.009, 1.293 and 0.430 respectively. (It is accepted that the accuracy of production estimations is dependent on the shortness of the sampling interval, (Zaika, 1973)). An alternative method of calculating production using the weight- specific growth-rate (method 3A, Crisp, 1971; Bergh, 1974) gave mean values of 2.328, 1.588 and 0.682 for the bay for 1975-6, which, given the constraint of sampling and analytical errors in this method, are in satisfactory agreement with the other values.

The annual production, as with growth, was influenced mainly by sediment physico-chemical factors and the tidal height. The highest production occurred in areas where a fine-sand/mud substrate occurred

Table 3 A comparison of Bertalanffy Growth Equations of *M. balthica* Populations.

| Area | Equation | $T_{95}$ (yrs) | Source |
|---|---|---|---|
| Torry Bay | $H_t=16.41(1-e^{-0.164(t+0.178)})$ | 18.9 | This study. |
| Wadden Sea | $H_t=18.43(1-e^{-0.514(t+0.500)})$ | 6.2 | From specimens obtained from J.J. Beukema (N.I.O.Z.) |
| Ythan Estuary | $H_t=22.18(1-e^{-0.185(t-0.492)})$ | 16.2 | Calculated from Chambers and Milne (1975) |
| Torry Bay | | 18.9 | This study. |
| North Sea (*1) | $L_t=28.30(1-e^{-0.112(t+0.577)})$ | 25.2 | Calculated from Vogel (1959) |
| Baltic Sea (*2) | $L_t=23.97(1-e^{-0.304(t-0.200)})$ | 10.3 | Calculated from Vogel (1959) |
| Lynher Estuary | $L_t=12.89(1-e^{-0.823(t-0.394)})$ | 4.0 | Calculated from Warwick and Price (1975) |
| Hudson Bay (*2) | $L_t=37.3(1-e^{-0.086(t-t_o)})$ | 34(*4) | Green (1973) |
| Hudson Bay (*3) | $L_t=21.1(1-e^{-0.309(t-t_o)})$ | 10(*4) | Green (1973) |

(*1 = sublittoral area; *2 = at M.L.W.; *3 = at 1m above M.L.W.; *4 = maximum age; $T_{95}$ - see Table 2; $H_t$ = shell height at time t (yrs); $L_t$ = shell length at time t (yrs).)

at a mid-tidal level and, as expected, in areas which supported the greatest biomass. The productivity increased with a decrease in both particle size and tidal height. This variation was ultimately associated with food quality, quantity and availability.

A comparison of absolute production and biomass levels with other areas is of little value because of the large spatial variation observed within a single area. However, a comparison of productivity is of greater value (Table 5). While some areas of Torry Bay supported population biomasses higher than elsewhere, the productivity of that biomass was clearly lower than other intertidal areas and more in agreement with sublittoral areas (although intertidal variation can be as large as geographical variation). In comparison with the P/B values of 0.35 - 0.49 calculated for Torry Bay, intertidal temperate populations of *M. balthica* had generally P/B values of *c* 1.0 (exceptionally to 2) while sublittoral and/or arctic populations have values <0.5. Ankar (1980), however, studied a sublittoral, arctic population and obtained a P/B of 4; this anomalously high figure was attributable to an unusual population age-structure.

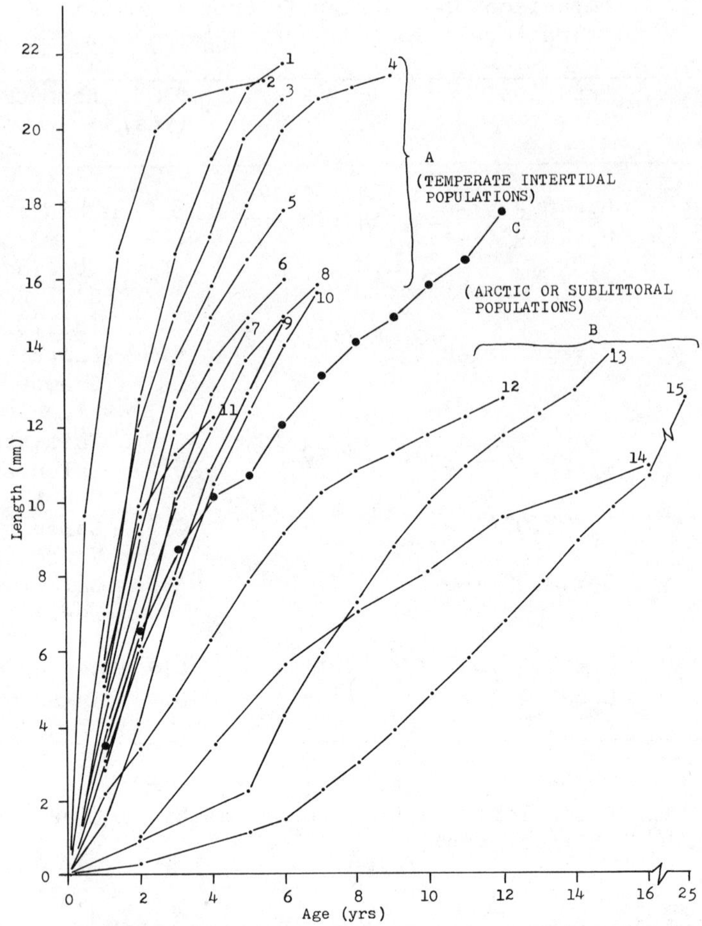


Fig. 4 The Geographical Variation in M. Balthica growth. Legend: C= the Torry Bay population; 1 Wadden Sea (specimens from N.I.O.Z.); 2 Rand Harbour, Massachusetts (Gilbert, 1973); 3 Wadden Sea (Lammens, 1967); 4 North Sea (Vogel, 1959); 5 Hudsons Bay: 1.1m above MLW (Green, 1973); 6 Kiel Bay (Vogel, 1959); 7 Ythan Estuary (Chambers and Milne, 1975); 8 Gulf of Finland (Vogel, 1959); 9 Hudsons Bay, at MLW (Green, 1959); 10 Gulf of Finland, 3m (Segestrale, 1960); 11 Lynher Estuary (Warwick and Price, 1975); 12 St. Lawrence Estuary (Lavoie et al, 1968); 13 Gulf of Finland, 20m (Segestrale, 1960); 14 White Sea (Semenova, 1970); 15 Gulf of Finland, 35m (Segestrale, 1960).
Nos. 2-6, 8-10, 12-15 inclusive from Gilbert (1973).

Table 4 The Production (g flesh weight $m^{-2}yr^{-1}$), Mean Biomass (g fdw $m^{-2}$) and Annual P/B of M. Balthica at Each Station.

| Station | 1975-6 B | P | P/B | 1976-7 B | P | P/B |
|---|---|---|---|---|---|---|
| A1 | 7.571 | 3.429 | 0.453 | 2.248 | 2.474 | 1.108 |
| A2 | 1.415 | 0.564 | 0.398 | 0.508 | 0.194 | 0.382 |
| A3 | 2.101 | 0.812 | 0.386 | 0.710 | 0.267 | 0.376 |
| A4 | 1.869 | 0.669 | 0.358 | 0.948 | 0.350 | 0.369 |
| B1 | 1.258 | 0.614 | 0.488 | 2.242 | 0.964 | 0.430 |
| B2 | 5.242 | 2.554 | 0.487 | 5.692 | 2.550 | 0.428 |
| B3 | 9.095 | 4.395 | 0.483 | 3.009 | 1.293 | 0.430 |
| B4 | 3.127 | 0.521 | 0.167 | 2.464 | 0.883 | 0.358 |
| B5 | 4.084 | 0.655 | 0.160 | 3.013 | 1.098 | 0.364 |
| C1 | 1.834 | 0.795 | 0.433 | 1.358 | 1.436 | 1.058 |
| C2 | 1.076 | 0.219 | 0.204 | 0.819 | 0.348 | 0.425 |
| C3 | 1.961 | 0.350 | 0.179 | 0.923 | 0.378 | 0.409 |
| C4 | 3.212 | 0.531 | 0.165 | 1.636 | 0.586 | 0.358 |
| C5 | 0.820 | -0.290 | -0.354 | 0.481 | 0.056 | 0.116 |

These data (and others not reported here) indicate that when compared to other similar areas the M. balthica population studied was under stress as indicated by its anomalous production ecology notably a reduced productivity; that stress cannot be explained by the environmental variables studied but may be caused by environmental abiotic factors. Of these factors the level of pollution is of most concern in view of the known levels of heavy metal and other pollutant accumulation (e.g. Fig. 1). It is of note that although the growth and production of many M. balthica populations have been estimated, in comparison with the Forth estuary those populations were in unpolluted areas.

## AN ASSESSMENT OF BIOTIC POLLUTION-EFFECTS STUDIES

While there has been much research on the elevation or depression of microbial productivity due to pollution stress (e.g. Mathews et al, 1982), few studies have been performed on the use of macrobiotic production ecology parameters in pollution studies. Both Leppakoski (1975) and Dauer and Conner (1980) found a change with pollution in the size frequency distribution of polychaetes; there was an absence of the smaller, less-tolerant life stages in the most polluted condtions. The latter authors suggest that an assessment of productivity in addition to community structure and a study of the species' mode of reproduction will reveal moderate pollution. However, these studies were concerned with organic pollution; it is widely

Table 5. <u>M. balthica</u> somatic production and productivity. (Range of values in parentheses.)

| Situation | Location | Annual $\overline{B}$ (g fdw $m^{-2}$) | Annual $\overline{P}$ (g fdw $m^{-2}$) | Annual $P/\overline{B}$ | Author |
|---|---|---|---|---|---|
| Intertidal | Forth Estuary | 3.190 (0.8 - 9.0) | 1.130 (-0.29 - 4.4) | 0.354 | This Study (1975/76 Data). |
| | Forth Estuary | 1.880 (0.48 - 6.0) | 0.920 (0.055 - 2.5) | 0.489 | This Study (1976/77 Data). |
| | Wadden Sea | 3.5 | 2.5 | 0.7 | Beukema (1971). |
| | Nova Scotia, Estuary | 1.26 | 1.93 | 1.53 | Burke and Mann (1974). |
| | Ythan Estuary | 2.77 | 5.73 10.07 | 2.07 | Chambers and Milne (1975b). |
| | Lynher Estuary | 0.337 | 0.308 | 0.9 | Warwick and Price (1975). |
| | Grevelingen Estuary | (0.01 - 2.93) | (-0.74 - 3.40) | 0.96 (-0.25 - 1.93) | Wolff and De Wolf (1977). |
| Sublittoral | Gulf of Finland | 8.97 | 4.22 | 0.47 | Segestråle (1960) by Bergh (1974). |
| | S. Baltic Sea | 2.60 | 1.01 | 0.388 | Bergh (1974) *[1]. |
| | N. Baltic Sea | 4.2 | 1.7 | 0.45 | Ankar and Elmgren (1976). |
| | N. Baltic Sea | 0.27 | 0.1 | 0.37 | Cederwall (1977). |

*[1] Calculated from Bergh (1974) using conversion figures given by Ankar (1976).

accepted that the total production of an area will increase with modest organic enrichment (Pearson and Rosenberg, 1978) but whether that is the case for either industrial pollution (as an increase of persistent pollutants) or productivity is unknown. Gray (1981), mentions briefly that chemical stress will reduce a individual's productivity and that monitoring by growth rate parameters possibly will be an extremely useful and sensitive method.

The experience of the *Macoma* case does indicate that pollution stress may result in anomalous production-ecology values or reduced productivity and suggests that production methods may detect sublethal effects of pollution. The present authors, in suggesting the use of production ecology methods acknowledge the many problems involved in calculating and interpreting macrofaunal growth and productivity, e.g. the inclusion or omission of production due to reproduction of all life stages or of ages treated separately (Hughes, 1970; Burke and Mann, 1974; Chambers and Milne, 1975). Possibly of greatest importance is that populations under comparison should have the same or a similar age structure relative to their life spans, since a decrease in P/B with increasing age is probably usual for most species (Ansell et al, 1978; Elliott, 1979; Ankar, 1980). In the use of these parameters for assessing the state of populations, methodologies would have to be standardised.

In detecting stress on estuarine invertebrates, the use of production (cf. physiological studies by Bayne and his co-workers). Stress attributable to non-organic pollution may increase respiration which would lead to decreased production. Therefore, a field-based estimation of respiration together with growth and production studies could parallel physiological ecology studies, (Bayne et al, 1977; de Wilde, 1973). The use of respiration measurements can also provide a shorter method than conventional methods in the estimation of production (McNeil and Lawton, 1970; Phillipson, 1970; Miller et al, 1971). However, in view of the possible inaccuracies in this method it is unknown whether variations in productivity due to stress would be detected.

Within the studies advocated here is the decision of which species to monitor. There is a wealth of data from community studies in pollution assessment (see below) on indicator species, i.e. those opportunistic, progressive and reselected species whose populations increase under polluted conditions. However, as with Gray (1979), the present authors suggest that it is species which neither thrive nor become severely reduced under polluted conditions that are most suitable for study in relation to pollution effects. These species would have stable populations, k-selected and/or T (tolerance) - selected strategies (as an adaptation to stress (Gray, 1979)) and be included in the second-order progressive and regressive species of Leppakoski (1975) and transitory species of Pearson and Rosenberg (1978). It is of note that recent literature on *M. balthica* makes it a suitable species for comparative and monitoring studies.

Any consideration of single species or communities affected by pollution (and environmental variability) should include acknowledgement of possible genetical and morphological variability, especially within species identified as pollution indicators (e.g. Rasmussen, 1973; Grassle and Grassle, 1977; together with personal observations).

The majority of techniques presently used to assess pollution effects and on which estuarine management decisions are made involve faunal community study. Indices of diversty (e.g. Shannon-Weiner, Shannon and Weaver, 1963), dominance and evenness (e.g. Pielou, 1966), rarefaction curves (Sanders, 1968), log-normal statistics (Preston, 1948, 1980) and classification methods of site association were all originally used to describe communities and have been used recently and extensively to study the structural response of an area to pollution (e.g. Leppakoski, 1975; Gray, 1981; etc.). Other methods, such as species/abundance/biomass (S.A.B.) curves, the benthic pollution index (B.P.I.), and the infaunal trophic index, have been used especially for ascertaining pollution effects along a spatial or temporal gradient (Pearson and Rosenberg, 1978; McLusky, 1982; Leppakoski, 1975; Bascom, 1982; respectively). The former, community-related methods, are more valid for use on homogeneous areas where the concept of the fauna being distributed as discrete communities or associations is accepted. In estuaries and close-shore areas, where the continua concept of species distribution ismore apt, where fauna change along gradients and where when normal environmental gradient or pollution gradient effects require to be studied, then the latter methods such as ordination, S.A.B., B.P.I., etc. should be used. Mills (1969), Hughes and Thomas (1971) and Gray (1981) and Herricks and Cairns (1982), review aquatic cmmunity structure analysis and its response under polluted conditions, while Hughes and Thomas (1971), Green (1980) and Jeffers (1978) give details of multivariate techniques useful in community and continual studies on benthic populations.

It is acknowledged that a major problem, with any method which detects species, population or community stress, is to separate stress arising from adverse physical or other natural conditions from that which is indicative of anthropogenic influences. In addition, the estuarine species encountered and used in monitoring studies will by definition be euryoecious and hence more tolerant of adverse or widely varying conditions.

The levels of pollutants in estuarine biota has been another area of intensive study; however, this does not often give an assessment of stress on the individuals. While there are a few studies on growth, etc. in high pollutant-level environments, most studies are concerned either with contamination, rather than stress and deleterious effects, or with lethal toxicities. Studies on detoxification of pollutants by organisms, the removal of large amounts of pollutants have been performed. (For reviews see Phillips, 1980; Moore, 1981). However, variation within the parameters of size, age, flesh condition, season, sex and, where necessary, shore position of animals used in accumulation studies appears to be a neglected area. The absence of much methodological standardisation invalidates statistical comparison between populations. In the Forth, and elsewhere, studies are now integrating the use of production ecology parameters (changes in growth and condition etc.), pollutant uptake using marked and caged animals and methodological standardisation. These studies parallel recent work on methodological standardisation in physiological stress measurements (Bayne et al, 1981).

To put the above consideration in the context of estuarine management, biota studies will lead to information on estuarine usage and the relative importance of and interrelationships between the various components of the system. This then leads to a consideration of the priorities for an estuarine area and the need to satisfy economic as well as ecological needs and ultimately to management decisions. A study of the biota will show a long-term integration of effects coupled with a relatively short-term response to a change in pollution (as shown by Pearson and Rosenberg, 1978; McLusky, 1982). On a local note, work has been completed or is in progress to assess the production of various parts of the Forth estuary and Firth to obtain an estimation of the relative importance of those parts. This is important in the light of industrial demands on the area for reclamation or effluent discharge. On a species level, work is in progress on single species production determination at varying distances from pollutant input.

These considerations are not to denigrate studies on community responses to pollution - in the Forth community indices are used as a first-step, rapid assessment of the health of an area adjacent to polluting discharges. The present authors concur with Mathews et al, (1982) who recommend an "holistic approach" to biomonitoring which includes the measurement of both structural and functional responses together with the need both to rationalise functional response assessment techniques and to distinguish natural environmental variability from individual, population, community or ecosystem response to stress. Although the technique of using production ecology to identify stress in marine and estuarine populations is in its infancy, initial data suggest that this may prove to be useful in biomonitoring.

## ACKNOWLEDGEMENTS

Our thanks to Dr. A.H. Griffiths of the F.R.P.B. Estuary Laboratories for providing the data for Fig. 1 and to our colleagues for helpful discussions. The Torry Bay study was financed by a University of Stirling Postgraduate Studentship to the first author.

## REFERENCES

Ankar, S. 1976. Final Report from the Benthic Macrofauna group. Contribution of the Asko Laboratory, University of Stockholm, No. 12.

Ankar, S., 1980. Growth and production of Macoma balthica in a Northern baltic soft bottom. Ophelia, supplement 1. 31-48.

Ankar, S. and Elmgren, R. 1976. The Benthic Macro- and Meio- fauna of the Asko-Landsort Area (Northern Baltic Proper). Contribution of the Asko Laboratory, No. 11, University of Stockholm.

Ansell, A.D., McLusky, D.S., Stirling, A., and Trevallion, A., 1978. Production and energy flow in the macrobenthos of two sandy beaches in South-West India. Proceedings of the Royal Society of Edinburgh 76 B (4) 269-296.

Bascom, W. 1982. The effects of waste disposal on the coastal waters of Southern California. Environmental Science Technology, 16 (4) 226-236A.

Bayne, B.L. 1978. The potential of bivalve molluscs for monitoring the effects of pollution. International Council for the exploration of the Sea, Rept. No. C.M. 1978/E¦ 43 pp 20.

Bayne, B.L., Widdows, J. and Newell, R.I.E. 1977. Physiological measurements on estuarine bivalve molluscs in the field. In¦ pp 56-68, Biology of Benthic Organisms, B.F. Keegan et al, (Eds), Pergamon Press, Oxford.

Bayne, B.L., Clarke, K.R. and Moore, M.M. 1981. Some practical considerations in the measurement of pollution effects on bivalve molluscs and some possible ecological consequences. Aquatic Toxicology 1, 159-174.

Bergh, H. 1974. Production of Macoma balthica (L) (Lamellibranchiata) and notes on other ecologically important aimals in Tvaren Bay in the Baltic. Zoon, 2, 143-152.

Beukema, J.J. 1971. The productivity of Macoma balthica (L) on a tidal flat area near Den Helder (Netherlands). Netherlands Journal of Zoology 21 (2), 213-4.

Beverton, R.J.M. and Holt, S.J., 1957. On the dynamics of exploited fish populations. Fishery Investigations, Series 2, London, 19, 1-533.

Burke, M.V. and Mann, K.H., 1974. Productivity and production¦ Biomass ratios of bivalve and gastropod populations in an eastern Canadian estuary. Journal of the Fishery Research Board, Canada, 31, 167-177.

Cederwall, H. 1977. Annual macrofauna production of a soft bottom in the Northern Baltic proper. p 155-164 in¦ Biology of Benthic Organisms, Keegan B.F. et al, (Eds), Pergamon Press, Oxford.

Chambers, M.R. and Milne, H. 1975. The production of Macoma balthica (L) in the Ythan estuary. Estuarine and Coastal Marine Science, 3, 443-455.

Crisp, D.J. 1971. Energy flow measurements. p. 197-279, Methods for the Study of Marine Benthos. Holme, N.A. and McIntyre, A.D. (Eds), I.B.P. Handbook No. 16, Blackwell Scientific Publ., Oxford.

Dauer, D.M. and Connor, W.G. 1980. Effects of moderate sewage input on benthic polychaete populations. Estuarine and Coastal Marine Science, 10, 335-346.

Elliott, M. 1979. Studies on the production ecology of several mollusc species in the estuarine Firth of Forth. Unpublished Ph.D. thesis, University of Stirling, pp. 430.

Gilbert, M.A. 1973. Growth rate, longevity and maximum size of Macoma balthica (L). Biological Bulletin, Marine biological laboratory, Woods Hole. 145, 119-126.

Grassle, J.F. and Grassle, J.P. 1977, Temporal adaptions in sibling species of Capitella, p. 177-190. In: Ecology of Marine Benthos, B.C. Coull (Ed.), Belle W. Baruch Library in Marine Science, No. 6. University of S. Carolina Press, Columbia.

Gray, J.S. 1979. Pollution-induced changes in populations. Philosophical Transactions of the Royal Society, London B 286, 545-61.

Gray, J.S. 1981. The Ecology of Marine Sediments. Cambridge University Press, Cambridge.

Green, R.H., 1973. Growth and mortality in an arctic intertidal population of Macoma balthica (Pelecypoda, Tellinidae). Journal of the Fisheries Research Board, Canada 30 1345-1348.

Green, R.H. 1980. The assessment of ecologic similarity. Annual Review of Ecological Systems, of Ecological Systems, 11, 1-14.

Herricks, E.E. and Cairns, J. 1982. Biological monitoring, Part III - receiving system methodology based on community structure. Water Research, 16 (2), 141-153.

Hughes, R.N. 1970. An energy budget for a tidal flat population of the bivalve, Scrobicularia plana (Da Costa). Journal of Animal Ecology, 39, 357-381.

Hughes, R.N. and Thomas, M.L.H., 1971. The classification and ordination of shallow-water benthic samples from Prince Edward Is., Canada. Journal of Experimental Marine Biology and Ecology, 7, 1-39.

Jeffers, J.N.R. 1978. An Introduction to Systems Analysis: with ecological applications. Edward Arnold, London.

Leppakoski, E., 1975. An assessment of the degree of pollution on the basis of macrozoobenthos in marine and brackish-water environments. Acta Academiae Aboensis, Series B., 35 (2), 1-96.

McErlean, A.J. 1964. Characteristics of Macoma balthica populations in the middle Patuxent Estuary. Chesapeake Science, 5 (4), 200-208.

McLusky, D.S. 1982. The impact of petrochemical effluents on the fauna of a intertidal estuarine mudflat. Estuarine and Coastal Shelf Science, 14 (5) 489-499.

McLusky, D.S. and Allan, D.G. 1976. Aspects of the biology of Macoma balthica (L) from the estuarine Firth of Forth. Journal of Molluscan Studies, 42, 31-45.

McLusky, D.S. and Elliott, M. 1981. The feeding and survival strategies of estuarine molluscs. p. 109-121 in: Feeding and Survival Strategies of Estuarine Organisms, N.V. Jones and W.J. Wolff, (Eds) Plenum Press, New York and London.

McLusky, D.S., Bryant, D.M., Elliott, M., Teare, M. and Moffat, G. 1976 Intertidal fauna of the industrialised Forth Estuary. Marine Pollution Bulletin 7, 48-51.

McLusky, D.S., Elliott, M. and Warnes, J.M., 1978. The impact of pollution on the intertidal fana of the estuarine Firth of Forth. p. 203-210 in¦ Physiology and Behaviour of Marine Organisms. D.S. McLusky and A.J. Berry, (Eds) Pergamon Press London.

McNeil, S. and Lawton, J.H., 1970. Annual production and respiration in animal populations. Nature, London 225, 472-474.

Mathews, R.A., Buikema, A.L., Cairns, J. and Rodgers, J.H. 1982. Biological monitoring, Part IIA - Receiving system functional methods, relationships and indices. Water Research, 16(2), 129-139.

Miller, R.J., Mann, K.H. and Scarratt, D.J., 1971. Production potential of a seaweed-lobster community in Eastern Canada. Journal of the Fisheries Research Board, Canada, 28(11), 1733-1738.

Mills, E.L. 1969. The community concept in marine zoology¦ a review. Journal of the Fisheries Research Board, Canada, 26, 1415-1428.

Moore, M.N. 1981. Elemental accumulation in organisms and food chains, p. 535-569 in¦ Analysis of Marine Ecosystems, A.R. Longhurst, (Ed) Academic Press, London.

Pearson, T.H. and Rosenberg, R. 1978. Macrobenthic succession in relation to organic enrichment and pollution. Oceaography and Marine Biology Annual Review 16, 229-311.

Phillips, D.J.H. 1980. Quantitative Aquatic Biological Indicators. Applied Science Publishers, Barking, Essex.

Phillipson, J. 1970. The 'best-estimate' of respiratory metabolism¦ its applicability to field situations. Polskie Achium Hydrobiologie 17, 31-41.

Pielou, E.C. 1966. The measurement of diversity in different types of biological conditions. Journal of Theoretical Biology, 13, 131-44.

Preston, F.W. 1948. The commonness and rarity of species. Ecology, 29, 254-83.

Preston, F.W. 1980. Noncanonical distributions of commonness and rarity. Ecology, 61, 88-97.

Rasmussen, E. 1973. Systematics and Ecology of Isefjord marine fauna. Ophelia 11 Nos. 1 and 2., p. 1-495.

Robertson, A.I. 1979. Relationship between annual production¦ biomass ratios and life span for marine maacrobenthos. Oecologia, (Berlin), 38, 193-202.

Sanders, H.L. 1968. Marine benthic diversity: a comparative study. American Naturalist 102, 243-282.

Segestrale, S.G. 1960. Investigations on Baltic populations of the bivalve Macoma balthica (L). I. Commentationes Biologicae Societas Scientatarum Fennica,. 23, 1-72.

Shannon, C.E. and Weaver, W.W., 1963. The Mathematical Theory of Communications, University of Illinois Press, Urbana, Illinois.

Vogel, K. 1959. Wachstumsunterbrechungen bei Lamellibranchiata und Brachiopoda. Neues Jahrbuch fur Geologie und Palaeontologie, Abhandlung B, B, 109, 109-129.

Warwick, R.M. and Price, R., 1975. Production of the macrofauna of an intertidal mudflat. Journal of the Marine Biological Association of the United Kingdom, 58 (1), 1-18.

Wilde, P.A.W.J. de 1973. A continuous flow apparatus for respiration recording of Macoma balthica, Netherlands Journal of Sea Research, 6, 157-162.

Wolff, W.J. and Wolf, L. de, 1977. Biomass and production of Zoobentos in the Grevelingen Estuary, the Netherlands. Estuarine and Coastal Marine Science 5, 1-24.

Zaika, V.E. 1973. Specific Production of Aquatic Invertebrates. Translated from Russian, I.P.S.T., Jerusalem and London.

# SHORT- AND LONG-TERM BIOSEDIMENTARY EVOLUTION OF TWO MACOMA BALTHICA COMMUNITIES

J.-P. Ducrotoy[1], M. Deprez[2] and J.-P. Dupont[2]

1.Station d'Etudes en Baie de Somme
115, quai Jeanne d'Arc
80 230 Saint Valery Sur Somme
France
2.Laboratoire de Biosedimentologie
10, boulevard de Broglie
76 130 Mont Saint Aignan
France

ABSTRACT

This paper investigates the influence of physical structures such as seawalls on the substratum and the associated *Macoma* communities. The breaching of a seawall in the Seine estuary induced a rapid change in sediment and macrozoobenthos. The drop in fine particles was followed by a reduction in both species diversity and numbers.

In contrast, the building of a new seawall in the Somme estuary was followed by a slower change in sediment and fauna. Siltation was greatest at high shore level.

Sediment changes were correlated with changes in the *Macoma balthica* population according to the proportion of fine particles in the substratum. In the Seine, the $0^+$ class accounted for 60-80% of the population in a sediment with 40-60% of silt, but only for 30-40% in a sediment with 20-30% silt. In the Somme, the silt increase was followed by a 3+ and 4+ year class decrease and a failure in the 0+ year class.

The growth of all age classes in the Seine was slowed by the hydrodynamic and sediment changes and the 0+ year class disappeared after six months. Growth in the Somme was better for newly recruited spat, but worse for the bigger animals.

Finally a change in the shell outline (H/L) of the older individuals was found with the increased siltation in the Somme.

## INTRODUCTION

The impact of human activities on estuarine ecosystems evolution has many aspects. There are discharges of synthetic or organic substances as well as destruction of the natural environment. It is this latter topic this paper wishes to investigate by assessing the influence of physical structures such as seawalls on the animal communities inhabiting intertidal soft substrate.

Several workers, notably Beukema (1974, 1976, 1981a, 1981b), have recommended the study of macrozoobenthos (i.e. fauna larger than 1mm) as a means of understanding the function of intertidal communities. Measurement of physical and chemical parameters, in the interstitial substrate or even in sea-water, presents some technical difficulties. On the other hand, the study of fauna, together with a study of the sediment, is a good alternative and provides a useful perspective to the biologist.

Desprez (1981) has shown, using mathematical models (dendrogramms, multi-factorial analysis), that sediment texture governs the intertidal macrozoobenthos distribution in the Seine estuary. In this way, bio-sedimentary studies are able to integrate the action of the whole of chemical and physical factors, under the assumption that the fauna distribution and the sediment evolution result from the hydrodynamic conditions.

The present work is part of a long-term programm of the Groupe d'Etude des Milieux Estuariens et Littoraux.

Two estuaries with seawalls, the Seine and the Somme, are compared. In the former, an opening was made in the embankment during the study period to let fishing-boats go through. It will be shown that the consequent perturbation was sudden, and the community evolution rapid. In the latter, the sedimentary character behind the embankment was not altered. A long-term interpretation of the phenomenon is suggested.

## LOCATION, MATERIAL AND METHODS

The development of the Seine estuary has been moulded by economic forces facilitating the industrial development of Le Havre and Rouen. The latest seawalls were necessary to obtain a permanent navigable channel. These embankments favoured the natural siltation of the estuary which had reduced the intertidal area to 30 km$^2$ by 1980 (Figures 1 and 2a). On both sides of the navigation channel, a

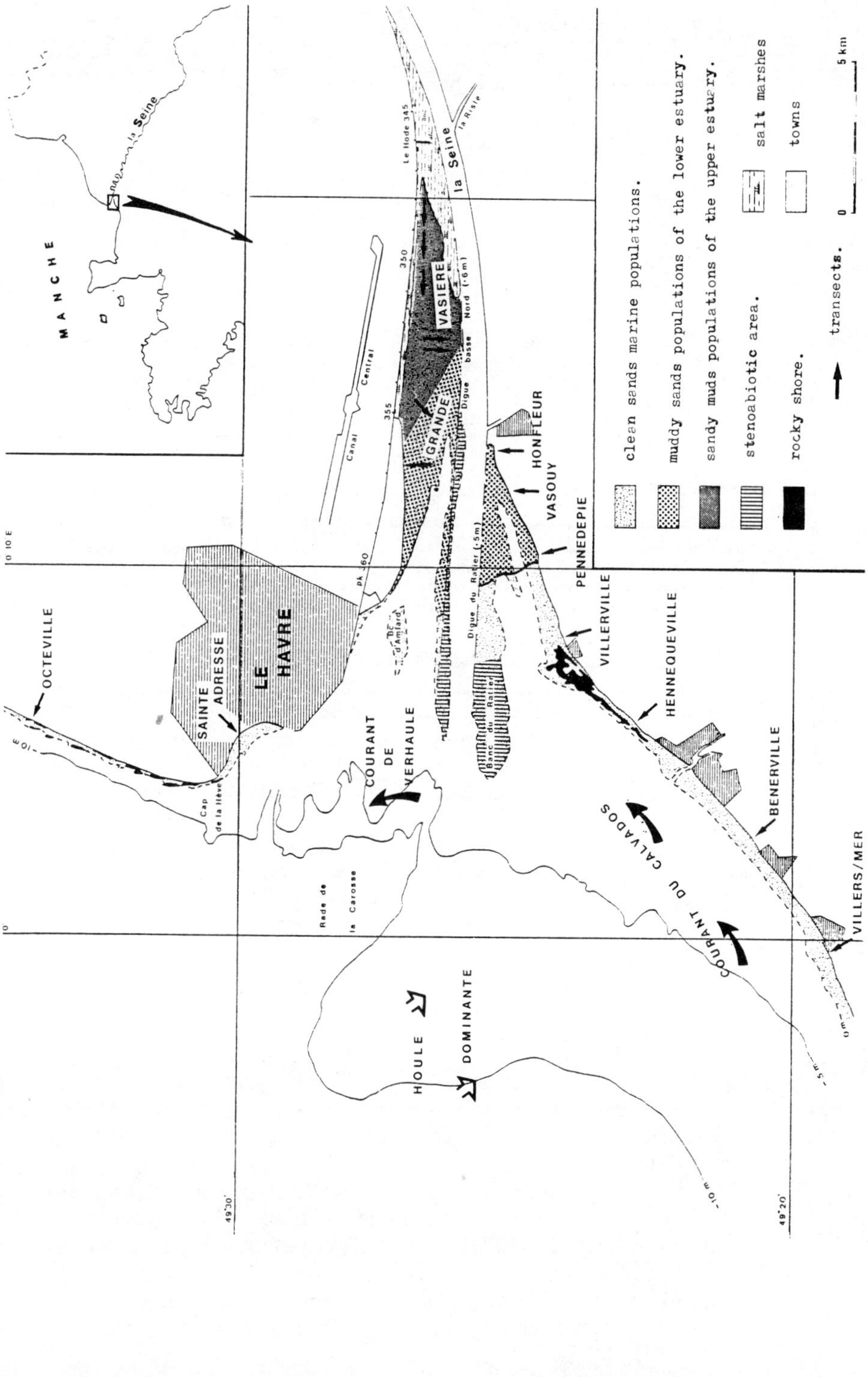


Fig. 1 Seine estuary. Inset: location in France. Solid arrows: currents. Light arrows: main swell direction.

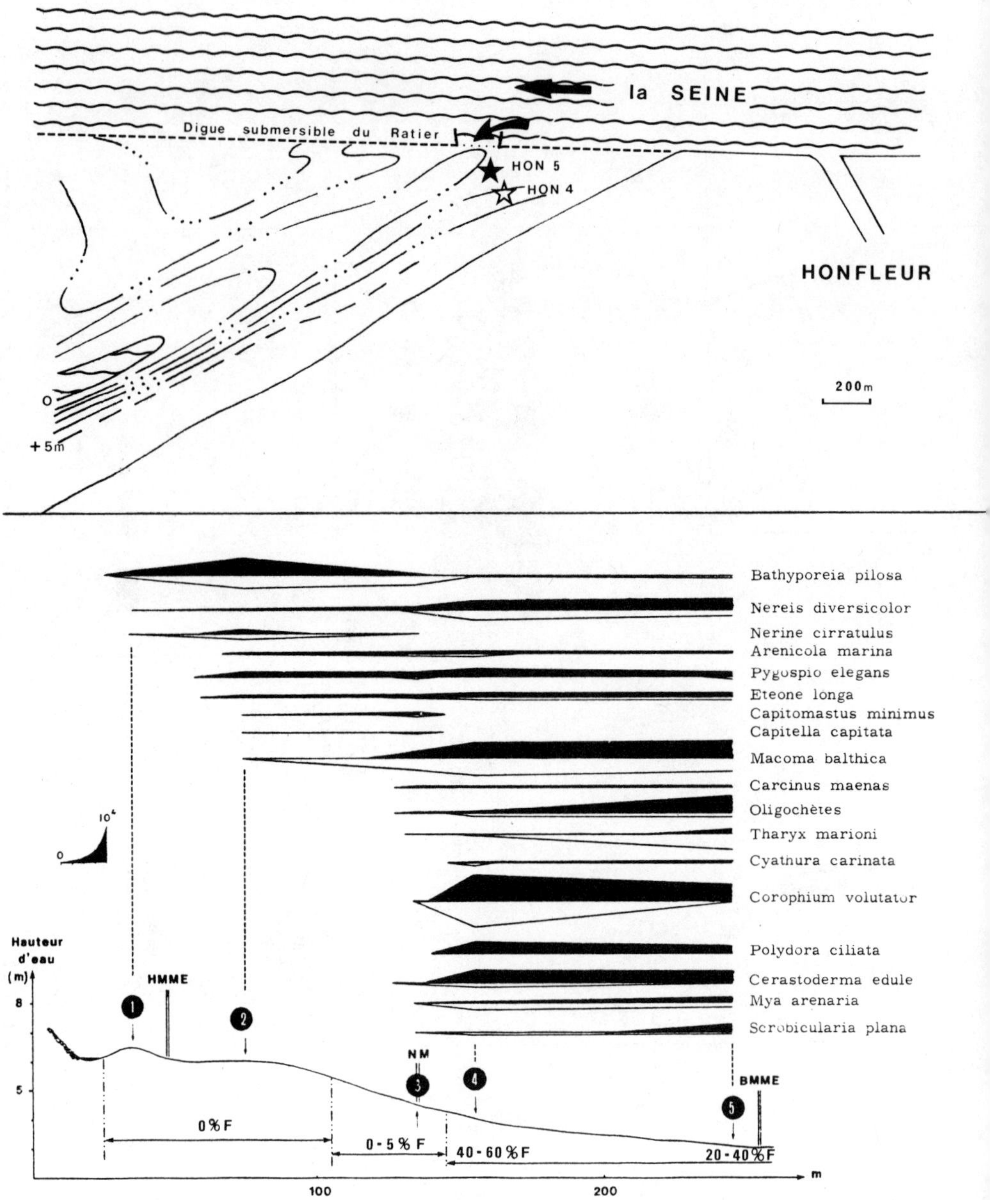


Fig. 2 a. Seine estuary, showing sea wall ("digue submersible") sample sites (stars) amd dyke opening (arrow).
b. Seine estuary, HON transect, showing distances and height above chart datum, (BMME= low water neap, NM= mid tide, HMME= high water neap) proportion of sediment fines (%F), numbers of stations sampled in 1978-81 (circled) and species numbers and distribution in winter (blank) and summer (solid) 1979.

flow-channel in the south and an ebb-channel in the north can still be seen as the last vestiges of the natural estuary. In the former are the two surveyed stations (Figure 2b), near a gap in the submersible embankment which was opened in July 1980.

Desprez (1981) described three intertidal communities in the Seine estuary which are characteristic of three sediment types:

- the marine community of clean sands,
- the lower estuary community of muddy sands,
- the upper estuary community of muds.

The Somme estuary in the "Baie de Somme" belongs to the estuarine complex comprising the Picard Maritime Plain where several small coastal rivers flow into the Channel. The Somme and the Maye estuaries form the "Baie de Somme" and total some 70 $km^2$ of intertidal area. An important factor is tidal range which reaches 9,80 m above zero level of French marine charts. Simon et al (1981) described the benthic communities belonging to the soft substrate of the Somme estuary. They linked this information with the sediment characteristics of the substrate (Figure 3) and were able to describe six "faciès":

- very fine sand with _Corophium volutator_,
- firm muddy sand with _Nereis diversicolor_,
- muddy sand where _Nereis diversicolor_ is dominant,
- sand with Spionidae,
- fine sand with _Arenicola marina_,
- muddy sand with _Cerastoderma edule_.

The Saint-Valery-sur-Somme seawall was erected in 1969 to canalize the Somme flow. This embankment favours the isolation of the South area (Hourdel-Hornu: HH) which was already sheltered between the Hourdel offshore bar and the Cretaceous shelf behind.

Two stations were selected in the Seine estuary for this study:

- _HON 5_:
  + 3m (above French marine charts zero level) in 1978,
  + 4m in 1981.
  muddy sand with _Macoma balthica_.
  the majority of the biosedimentary project concerns this station.

- _HON 4_:
  + 4m in 1978,
  + 4.5 m in 1981.
  muddy sand with _Macoma baltica_, under a _Corophium volutator_ "facies" owing to a superficial mud layer.
  _Macoma balthica_ growth patterns were followed at this station.

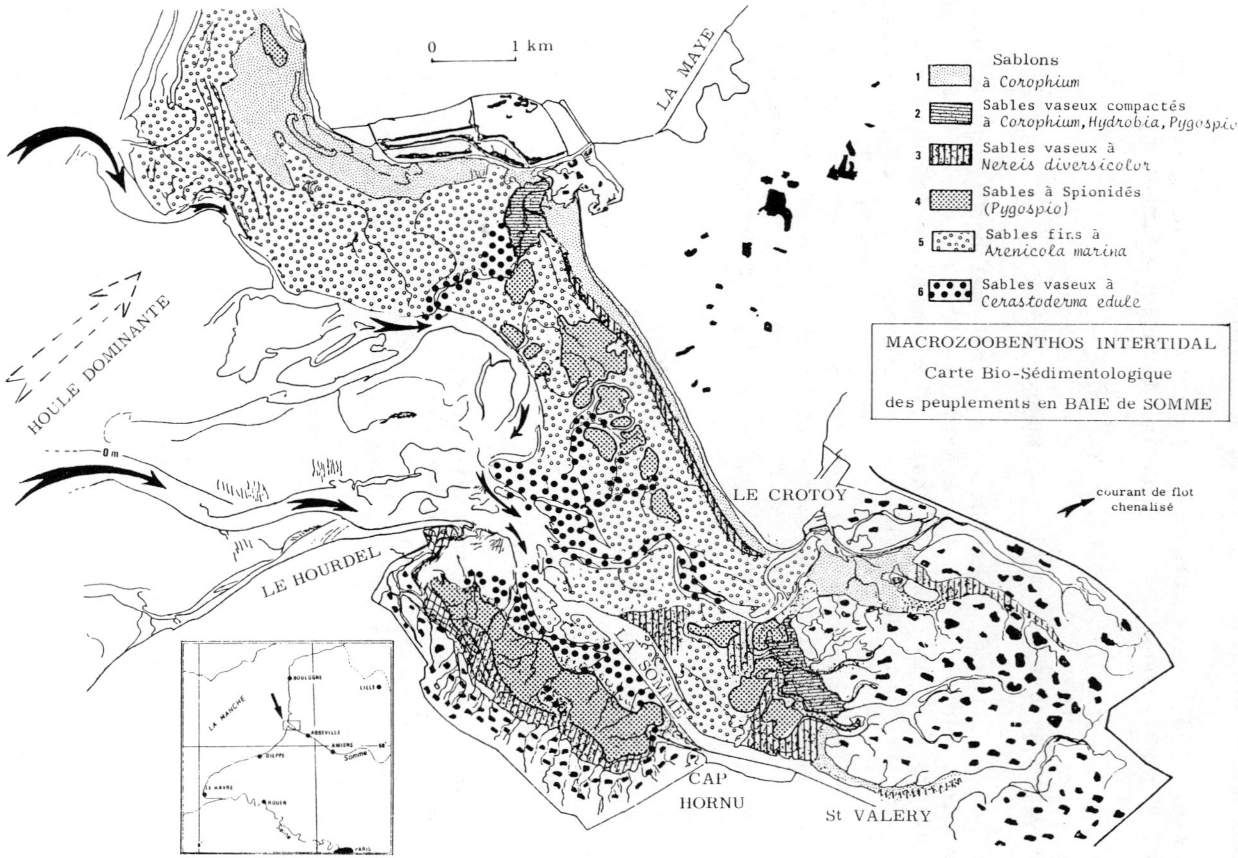


Fig. 3 Somme estuary. Notation as Fig 1, showing the six main communities in relation to sediment; (see text p110).

At HON 5, samples 1/8 $m^2$ in area (0.5 x 0.25 m) were dug out to a depth of 0.2 m.

At HON 4, the growth of Macoma balthica was surveyed from 1/10 $m^2$ samples (0.15 x 0.666 m) dug to a depth of 0.1 m.

Both stations were sampled at two month intervals.

In the Somme estuary, three 900 $m^2$ quadrats were delimited by permanent marks. These were in the south part of the estuary, behind the seawall, along a transect which has been continuously surveyed since 1978 (Figure 4):

- HH A:
  + 8 m (above French marine charts zero level) in 1981.
  muddy sand with Nereis diversicolor.

- HH B:
  + 7.5 m in 1981.
  sands with Spionidae (Pygospio elegans)

- HH C:
  + 6.2 m in 1981
  muddy sand with Cerastoderma edule.

From each quadrat, ten 1/25 $m^2$ samples were taken with a cylindrical corer to a depth of 25 cm during the same tide period. Samples were distributed over the quadrat along rectangular co-ordinates selected by random number tables.

In both estuaries, sediments were sieved in the field through a 1 mm square mesh. Animals were kept in saline formalin and coloured with phloxin B. In the laboratory, they were identified and counted. The biometry project deals with bivalve shells and Macoma balthica in particular. The length, height and width of each shell was measured and the age of the animal was assessed from the number of growth bands (Chambers and Milne, 1975). Measurements were taken to the nearest 0.10 mm and pooled into 1 mm classes. For the analysis of these results in the Somme, the model of Bhattacharya (1967) was used. To check the results given by this method, several animals belonging to the peaks of the size frequency curve were chosen and the growth bands counted. A good correlation existed between the two methods. Together with every faunal sample, two substrate samples were taken for sediment analysis:

- one was used for inclusive granulometric spectra (on an AFNOR column for sands and by a micro-granulometric methods for the fine particles) and carbonate analysis;

- the second was used for organic matter analysis.

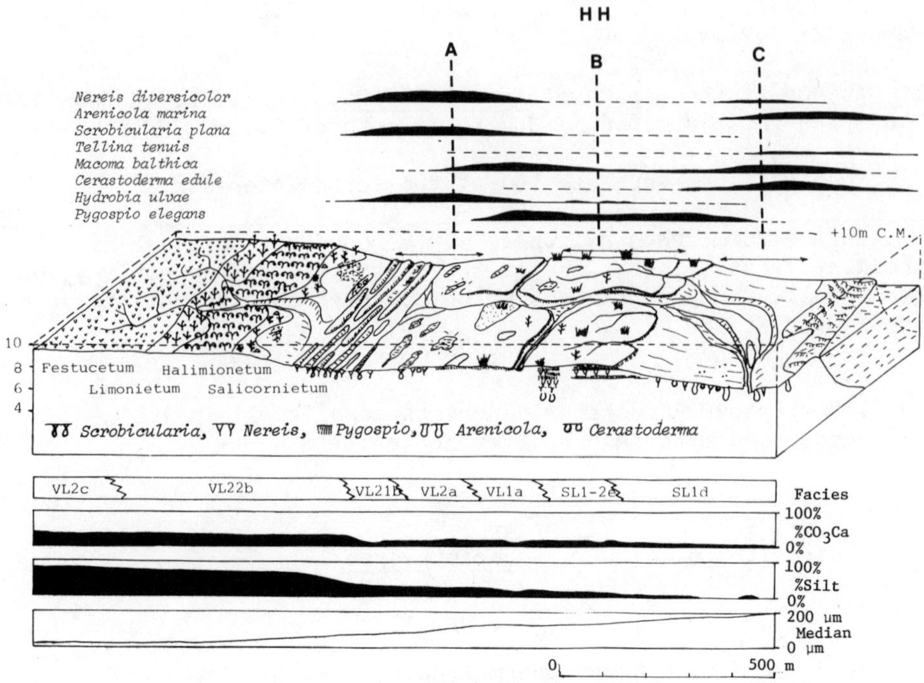


Fig. 4 Somme estuary: HH transect, showing (from foot) median grain size, % silt, % $CaCO_3$, facies (after Larsonneur, 1977), perspective view and height above chart datum of transect, and species numbers and distribution.

## RESULTS

In the Seine estuary, the opening of the seawall (digue du Ratier: 20 m long x 1.50 m deep) in July 1980 had a two-fold effect:

- sand was added from the sailing channel,
- silt was washed away.

At the HON 5 station, close to the seawall opening, the percentage of fine particles (less than 50 µm) dropped from a mean value of 32% to 7%; that is the sediments became less muddy. The opening has hardly altered the sand accretion in this area: the building-up has continued (1 m from 1978 to 1981) (Figure 5), as the station is situated in a discarded flow-channel.

The HON 4 station was characterized by a layer of mud (50-60% fine particles) above muddy sand similar to the sediment of HON 5 before the alteration. The HON 4 station is further from the seawall, and has not yet shown noticeable change which could be attributed to the opening of the dyke (Figure 6).

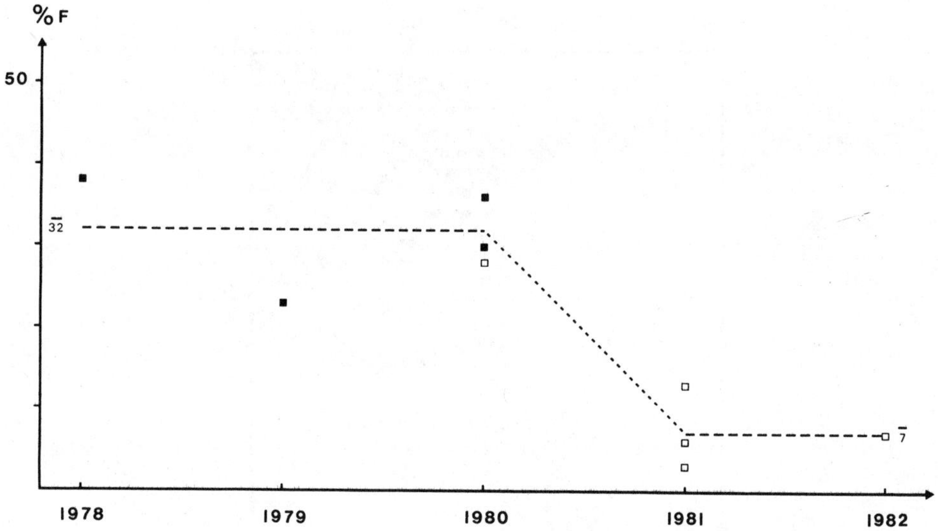


Fig. 5 Sediment change at site HON5 1978-82 (%F = percent <50 μm) before (■) and (□) after the sea wall opening

In the Somme estuary, the tendency has been towards the evolution of salt-marshes. The construction of the seawall in Saint-Valery, from the "Cap Hornu" was followed by a marked decrease in water currents and a stabilisation of the channels. Stabilisation of the mud-flats (because of the suppression of the Somme course shifting) permitted a massive increase in the population of Pygospio elegans and the establishment of tube-beds (Dupont, 1975). These structures permitted the deposition of suspended particles and the siltation of the whole studied area. The propagation of pioneer-plants accelerated this phenomenon towards the establishment of the salt-marsh.

Stations HH A, HH B, and HH C were arranged along an ascending granulometrical gradient, the sediment of which has a growing percentage of fine particles correlated with the construction of the seawall (Figures 7,8).

On the faunal side, the Seine estuary demonstrated short-term and the Somme estuary long-term community evolution.

The HON 5 population is representative of the low estuary community. It has a high species richness (fourteen species were collected throughout the project) with regard to the North side stations which belong to a oligospecific facies of the same community. This could be correlated with the low salinities in this area next to an ebb-channel.

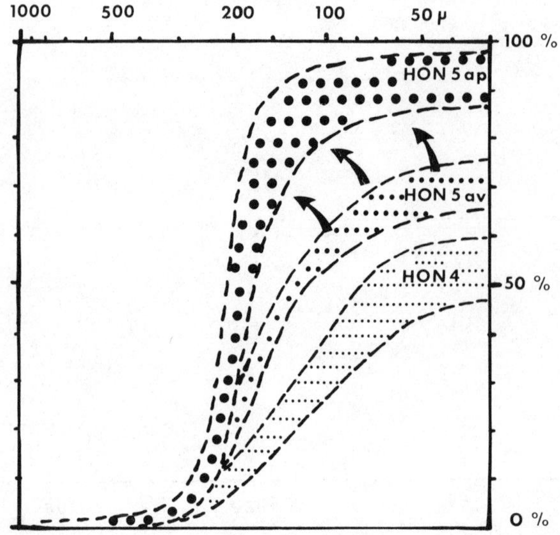


Fig. 6 Envelope curves of HON4 and HON5 stations. HON4 -1979 and 1982; HON5 av=1979, HON5 ap=1982. Arrows indicate direction of change at HON5 from 1979 to 1982.

The mean density of 4,837 individuals m$^{-2}$ was due, mainly, to oligochaetes (Tubificoides benedeni) and, secondly, to the polychaete Tharyx marioni and the bivalve Macoma balthica. The numbers of the first two species were related to the quantity of dead plant material in the sediment. The main associated species were Cerastoderma edule, Nereis diversicolor, Corophium volutator. This animal association was defined by Petersen (1913) as the .ul Macoma community.

After the opening of the seawall, the number of species decreased slightly; the mean number was 14.1 sp year$^{-1}$ before the opening (from July 1979 to June 1980) and 11 sp year$^{-1}$ after the opening (from April 1981 to March 1982) (Table 1). The new species which settled were characteristic of clean and slightly muddy sands namely, Spio filicornis, Bathyporeia sarsi and Nephthys cirrosa which colonized the marine transects, West of Pennedepie, and Capitomastus minimum which could also be found in the upper parts of the Honfleur transect (Figure 2b) and reached the lower stations. The densities decreased strongly from 4,814 individuals m$^{-2}$ year$^{-1}$ before 605 individuals m$^{-2}$ year$^{-1}$ after the opening (Table 1 and Figure 9). The faunal evolution was the direct consequence of the sediment evolution which in turn was the result of the new hydrodynamic conditions. The settlement of species characteristic of marine transects (Nephthys cirrosa, Mysella bidentata, Bathyporeia sarsi, Spio filicornis) provides an illustration of the prime importace of fine particles in limiting penetration of these species into the estuary. Further evi-

Table 1 Faunal changes at HON5 (Seine), showing average densities (Nos m$^{-2}$) before (July 1979 - June 1980) and after (April 1981 - March 1982) the opening of the sea wall.

| SPECIES | BEFORE | AFTER | CHANGE |
|---|---|---|---|
| a) Decreasing density species | | | |
| Pectinaria koreni | 275 | 0 | -100% |
| Polydora ciliata | 79 | 0 | -100% |
| Carcinus maenas | 50 | 0 | -100% |
| Arenicola marina | 13 | 0 | -100% |
| Cyathura carinata | 1 | 0 | -100% |
| Scrobicularia plana | 117 | 3 | -98% |
| Oligochaetes | 2,164 | 51 | -98% |
| Cerastoderma edule | 114 | 7 | -95% |
| Nereis diversicolor | 170 | 9 | -95% |
| Corophium volutator | 165 | 17 | -90% |
| Pholoe sp. | 7 | 1 | -86% |
| Macoma balthica | 848 | 232 | -73% |
| Tharyx marioni | 612 | 195 | -68% |
| Crangon crangon | 24 | 9 | -63% |
| Bathyporeia pilosa | 22 | 9 | -60% |
| Mya arenaria | 49 | 21 | -58% |
| Eteone longa | 39 | 17 | -57% |
| Nephthys hombergi | 18 | 9 | -50% |
| Pygospio elegans | 47 | 25 | -46% |
| Sub-total of individuals | 4,814 | 605 | -87% |
| No. of density decreasing species | 19 | 14 | |
| b) New species | | | |
| Nephthys cirrosa | 0 | 1 | |
| Mysella bidentata | 0 | 4 | |
| Bathyporeia sarsi | 0 | 5 | |
| Spio filicornis | 0 | 7 | |
| Capitomastus minimus | 0 | 16 | |
| Sub-total of individuals | 0 | 33 | |
| No. of new species | 0 | 5 | |
| Total No. of individuals | 4,814 | 638 | |

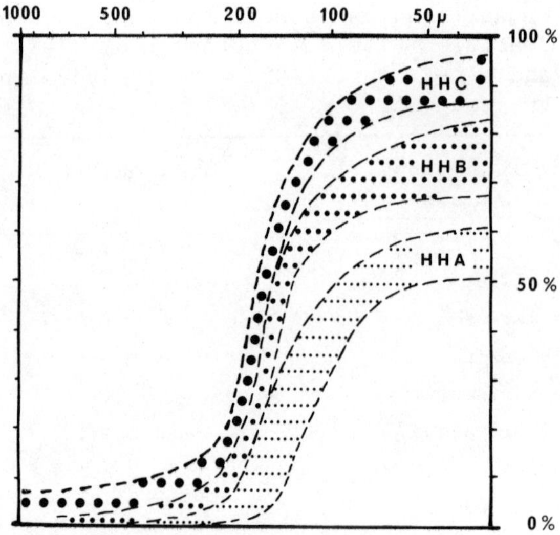


Fig. 7 Somme estuary sediment curves 1981 from high water to low water HHA, HHB, HHc.

dence for the limiting role of fine particles is provided by the downshore migration of Capitomastus minimus.

Sediment analysis in the Somme estuary showed a downshore progression of the salt-marsh along the HH transect. The amount of time spent on this project restricts the conclusions which can be drawn about the long-term evolution according to a strictly temporal pattern. On the other hand, it was possible to follow population succession along the transect. At any given point, conclusions may therefore be deduced about the time succession from the study of the spatial distribution.

Downshore, (HH C), the fauna diversity was low: 8 species year$^{-1}$ The two main species were bivalves: Cerastoderma edule (70-1,975 individuals m$^{-2}$), Macoma balthica (200-1,592 individuals m$^{-2}$).

(Numbers give minimum and maximum densities throughout an annual cycle). Another important species, because of its high numbers, but contributing little to the biomass was: Bathyporeia pilosa (185-879 individuals m$^{-2}$).

Further up, at HH B, the densities of these three species decreased: Cerastoderma edule (85-690 individuals m$^{-2}$), Macoma balthica (146-518 individuals m$^{-2}$), Bathyporeia pilosa was always absent from the samples.

Two species became important: Scrobicularia plana (10-45 individuals m$^{-2}$), Pygospio elegans (about 10,000 individuals m$^{-2}$, estimated by weight).

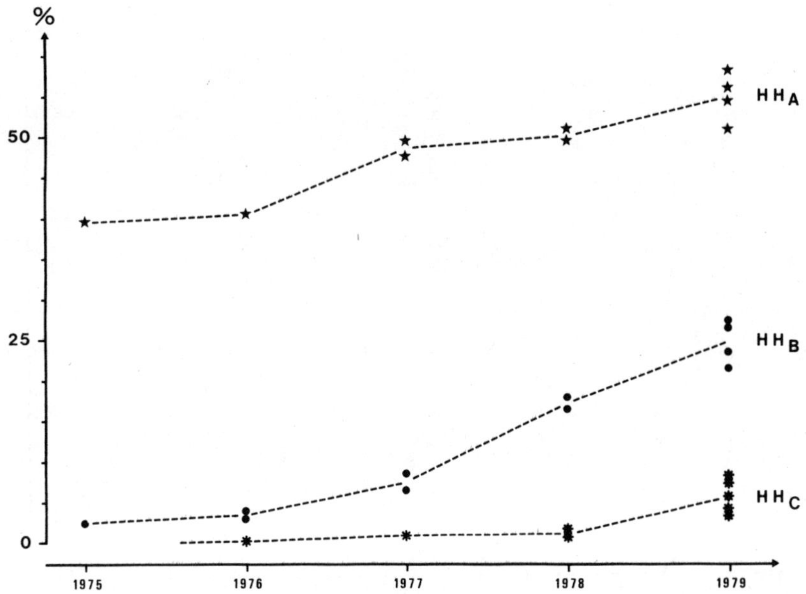


Fig. 8 HH transect 1975-1979 sediment change: % = percent fines (<50 μm).

Upshore (HH A), a higher diversity could be observed, due to the presence of low numbers of several species: Cerastoderma edule represented by scarce individuals throughout the samples has almost disappeared. and numbers of Pygospio elegans were low (1,500 ind $m^{-2}$). High densities were reached by: Scrobicularia plana (108-210 ind $m^{-2}$) Nereis diversicolor (1,690-4,698 ind $m^{-2}$) Hydrobia ulvae (more than 100,000 ind $m^{-2}$, estimated by weight).

Seasonally, Capitella capitata (up to 720 ind $m^{-2}$) could be observed Mya arenaria (3-5 ind $m^{-2}$) was fairly constantly distributed with low densities. Numbers of the ubiquitous species Macoma balthica (286-685 ind $m^{-2}$) remained steady (Figure 4).

The siltation following the construction of the seawall implies a chronological population series at a given point according to the space distribution which has been described.

In the long-run, it is logical to predict that the cockle-beds will die out to be replaced by Scrobicularia plana at HH C as the mud proportion increases. The result will be a similar facies to HH A leading eventually to a salt-marsh (Figure 10).

One of the principal aspects arising from the study of M. balthica was the large size of the Somme and Seine shells, up to and over 22 mm long. This measurement exceeds other published figures:

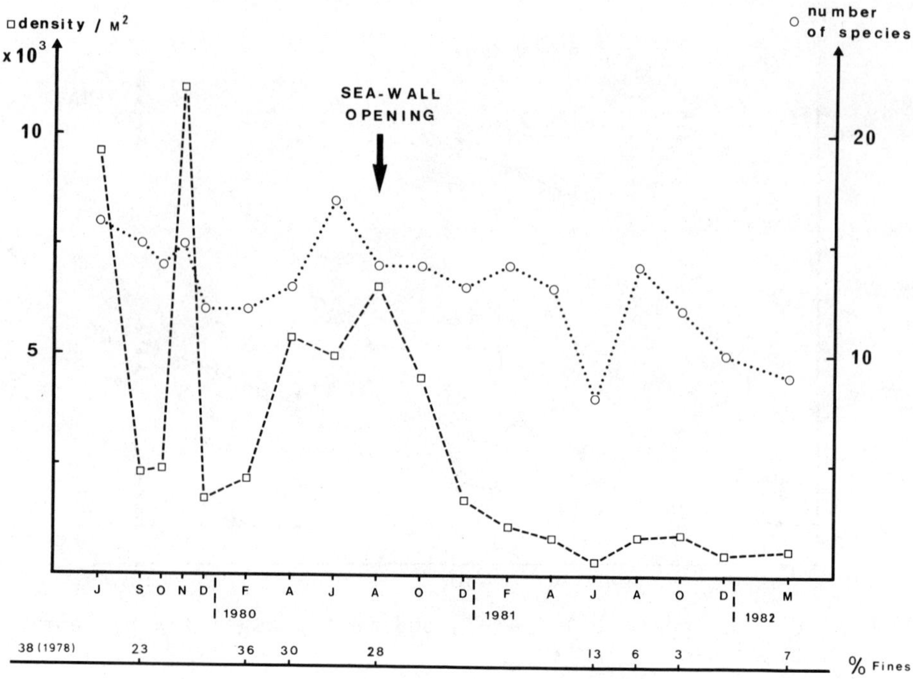


Fig. 9 Seine estuary HON5 site showing changes in density ($N^os\ m^{-2}$), species numbers, and %fines after this sea wall was breached (arrowed).

0–13 mm in Canada, 12 year-classes (Lavoie et al, 1968)
0–16 mm in Scotland, 6 year-classes (Chambers and Milne, 1975);
0–20 mm in England, (Reading, 1979);
0–21 mm in the Netherlands, 5 year-classes (Lammens, 1967);
0–16 mm in France (Gironde), 5 year-classes (Bachelot, 1980).

The impact of the opening of the dyke on the growth of M. balthica could be seen from the change in population structure and growth rate. The three main year-classes proportions (in per cent) were compared thus:

$0^+$: animals born the sampling year,
$1^+$: animals born the year preceeding sampling,
$2^+$...:older animals.

At the HON 4 and HON 5 stations, there was correlation between the percentage of fine particles in the sediment and the relative numbers of the $0^+$ class (Figure 11).

At the HON 5 station, the $0^+$ class accounted for 50-80% of the population, when the sediment contained more than 30% fine particles

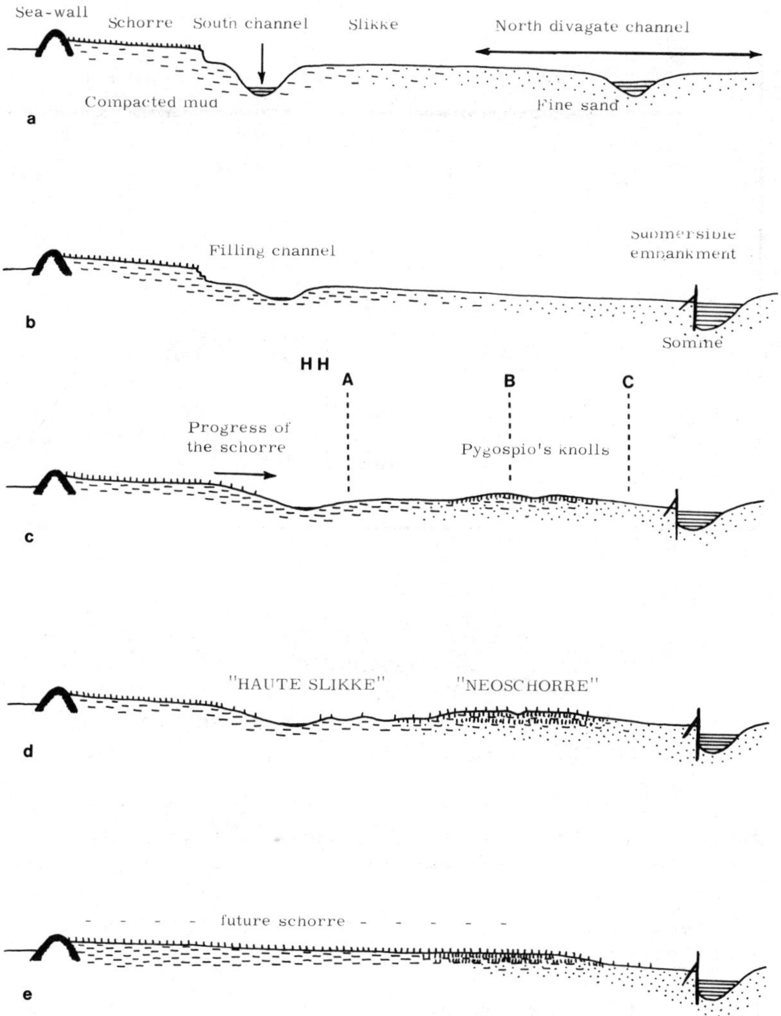


Fig. 10 Somme estuary change in HH transect: before (a) and after (b) sea wall construction, stabilisation of <u>Pygospio</u> facies (c), and future sequences (d,e) of mud flat (shlilcke) and salt marsh (shorre).

in 1979, but it represented only 30-40% when the sediment contained less than 30% of fine particles in 1980.

The lesser importance of the $0^+$ year-class was, largely, due to the seawall opening in July 1980 since the autumn maximum which characterised the month of November 1979 was not observed. The year 1981 corroborated the prime importance of the opening since the $0^+$ class accounted for a maximum of 17% of the population in August in a sediment with less than 10% fine particles.

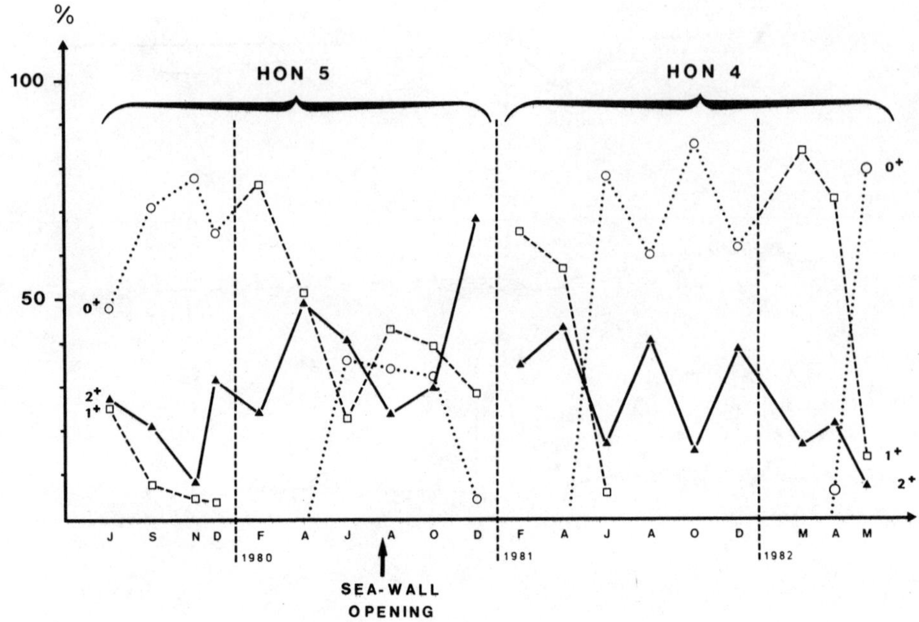


Fig. 11 M. balthica. Seasonal changes in $0^+$, $1^+$, $2^+$ year classes at HON4 (July 1979 - Dec 1980) and HON5 (Feb 1980 - May 1981). Note sea wall opening (arrowed) necessitating change in station.

At the HON 4 station, the $0^+$ year-class accounted for 60-80% of the population, in a muddy sediment with 40-60% fine particles.

The seawall opening had several consequences on the growth of Macoma balthica: (Figure 12a and 12b)

- the growth of each year-class slowed down, particularly in the young classes ($0^+$, $1^+$). Figures 12a and 12b show that growth was less disturbed by the opening at HON 4 than at HON 5, but the stress seemed less dependent on animal age at the former than at the second station. Growth became "normal" again in Spring 1981.

- Autumn recruitment was very low in 1980, and the spring cohort of the $0^+$ class, which settled from April to June, became scarce as early as December at HON 5, and the next June at HON 4.

- The $4^+$ and $5^+$ year classes disappeared from HON 5 in 1981. This was correlated with the big drop in density of Macoma balthica (Table 2).

- In 1981, the $0^+$ year-class was scarce at HON 5; the sediment evolution (especially the loss of fine particles) would appear

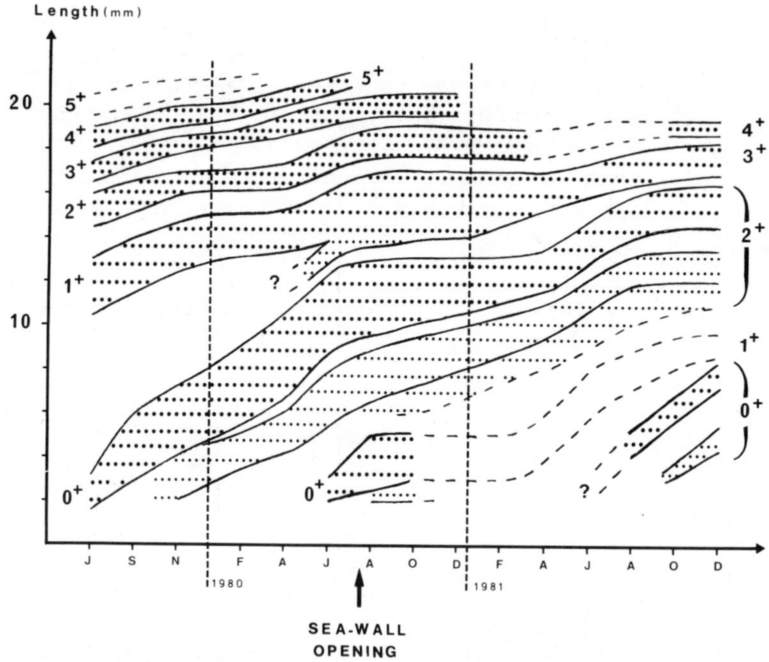


Fig. 12a M. balthica. Year classes at HON5 showing range of ages (mm) with age (growth rings), and change with the sea wall opening (arrowed).

to account for it since the spring recruitment was "normal" at HON 4 (78% of the population) where the sediment was not altered.

- The autumn cohort of 1979 remained distinct two years after its settlement, owing to the failure of the recruitment in 1980.

The population structure of M. balthica from the transect behind the seawall in the Somme (Table 2) shows that the animals are not distributed evenly, notably at HH A. Upshore, the substrate is not flat and exhibits irregularities due to a surface system of ebb-channels and pioneer plant settlement. Nevertheless, the data shows that for the three year-classes: $0^+$: animals born in 1981; animals born in 1980; $2^+$...: older animals:

- At HH A, spring and autumn recruitment never rise above 35% of the population and a great proportion of $1^+$ overwinter: 77.5% in March 1981.

- At HH B, autumn recruitment did not seem successful. On the other hand, the $0^+$ class account for 77% in June.

Table 2 M. balthica. Observed numbers (in brackets) and percentage in each year class in the Somme estuary, from higher to lower levels along the Hourdel-Hornu (HH) transect: HHA, HHB, HHC. S: time in weeks

| HH 1981 | | S 13<br>25<br>march | S 18<br>2<br>may | S 26<br>24<br>june | S 34<br>21<br>august | S 41<br>7<br>october |
|---|---|---|---|---|---|---|
| $0^+$ | A | (0) 0% | (1) 1% | (25) 25% | (98) 35% | (43) 30% |
| | B | (0) 0% | (1) 20% | (185) 77% | (17) 24% | (21) 14.5% |
| | C | (0) 0% | / | (54) 56% | (47) 47% | (254) 70% |
| $1^+$ | A | (45) 77.5% | (99) 72% | (42) 45% | (119) 42% | (47) 33% |
| | B | (16) 60% | (2) 40% | (42) 16% | (43) 62% | (111) 76.5% |
| | C | (19) 41.5% | / | (10) 10% | (15) 15% | (30) 8% |
| $2^+$ | A | (6) 10.5% | (11) 8% | (24) 25% | (59) 21% | (30) 21% |
| | B | (6) 26% | (1) 20% | (16) 7% | (2) 3% | (9) 6% |
| | C | (19) 41.5% | / | (14) 16% | (21) 21% | (51) 14% |
| $3^+$.. | A | (7) 12% | (28) 18% | (7) 8% | (7) 2% | (24) 16% |
| | B | (1) 14% | (1) 20% | (1) 1% | (8) 11% | (4) 3% |
| | C | (8) 17% | / | (16) 18% | (17) 17% | (30) 8% |

- Throughout the year, the proportion of $1^+$ is high at the high levels, especially at HH B.

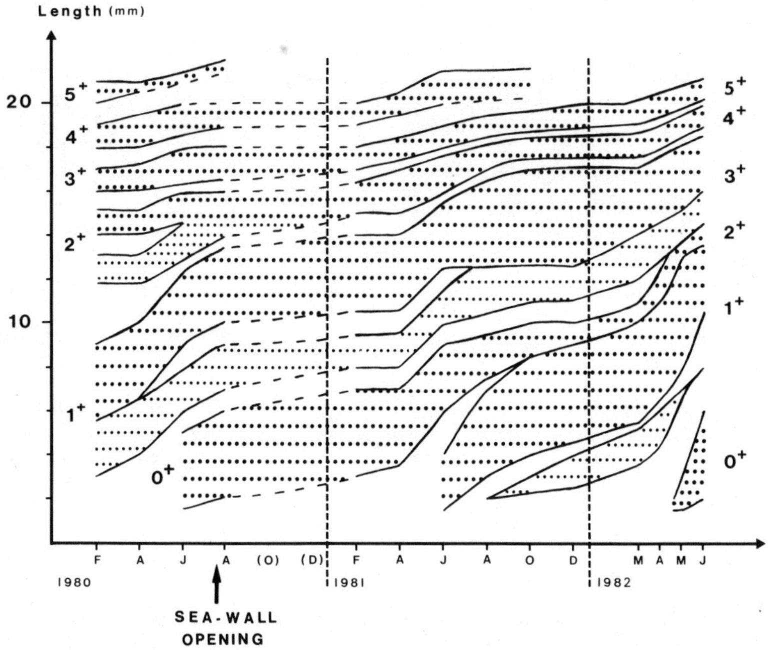


Fig. 12b M. balthica. Site HON4. Legend as Fig. 12a

- At HH C, recruitment is good in spring (56%) as well as in autumn (up to 70%).

Growth data for the different stations (Figure 13: age is estimated from observation of growth rings) showed no difference in growth rate nor in the shell length of the $1^+$ year-class. There was however some evidence to suggest growth was faster for the $0^+$ sampled at HH C. On the other hand, growth was almost nil for the animals by predation or because of a different ecolgical factor. The growth curves given by the Bhattacharya (1967) model show (Figure 14a, b, c):

- Good $0^+$ growth, although the recruitment was not successful at HH A. No shell was larger than 20 mm, confirming the pevious remark.

- Poor growth by the autumn cohort at HH B; the presence of $3^+$ and $4^+$ shells was noted.

- Bigger sizes were attained at HH C where a supplementary age class, $5^+$, was found.

In conclusion, the growth of young animals was good upshore in spite of a worse recruitment than downshore. The older age classes

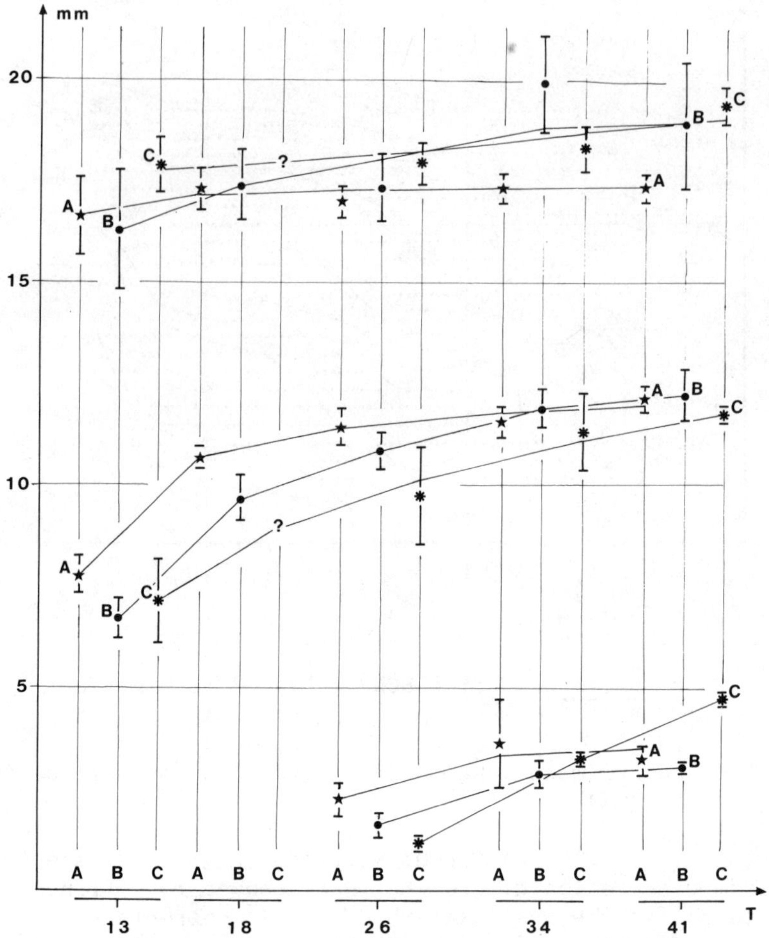


Fig. 13 M. balthica. Increase in shell length (L, mean ± 95% confidence limits) with time (T, weeks) at sites HHA (A), HHB (B) and HHC (C). ? indicates insufficient data.

disappear gradually with height up the shore from around mid tide level. Overall, a comparison of envelope-curves for the Seine and the Somme stations (Figures 6 and 7) show a similarity between: HON 4 and HH A, and especially between: HON 5 av and HH B; HON 5 ap and HH C. (See Figure 7 for explanation.)

If the sediment characteristics for the two estuaries are examined with respect to time, an inverse pattern of evolution can be seen (Figure 15). Comparison of the fauna from equivalent sediments in both areas reveals:

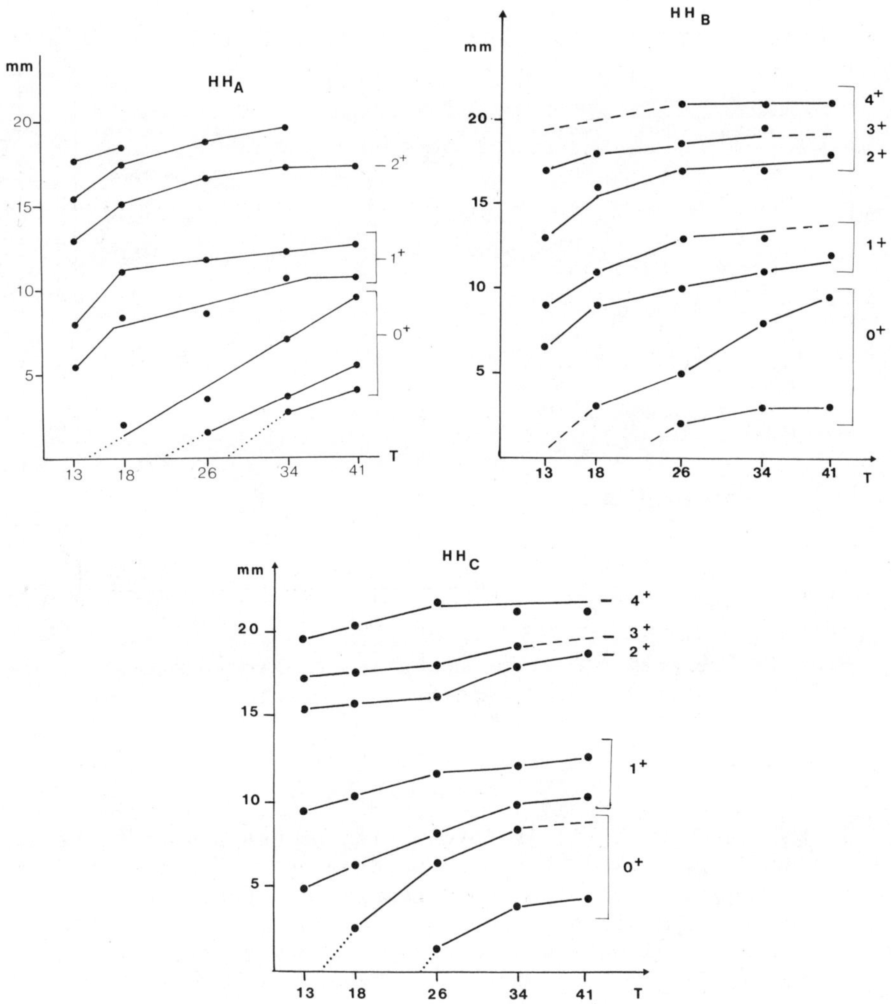


Fig. 14 *M. balthica* growth curves from Bhattacharya (1977) model at sites HHA (a), HHB (b), and HHC (c). Axes as Fig 13.

(1) species encountered in one or other estuary; for instance *Tharyx marioni* and *Corophium volutator* at Honfleur, and *Hydrobia ulvae* at Hourdel-Hornu.

(2) species designated as indicator-species. Grain size is the main factor influencing the numerical variation of these populations; for instance: *Scrobicularia plana* and *Nereis diversicolor*.

(3) other common species the increase in density of which is not necessarily linked with grain size; for instance: *Macoma balthica* and *Cerastoderma edule*.

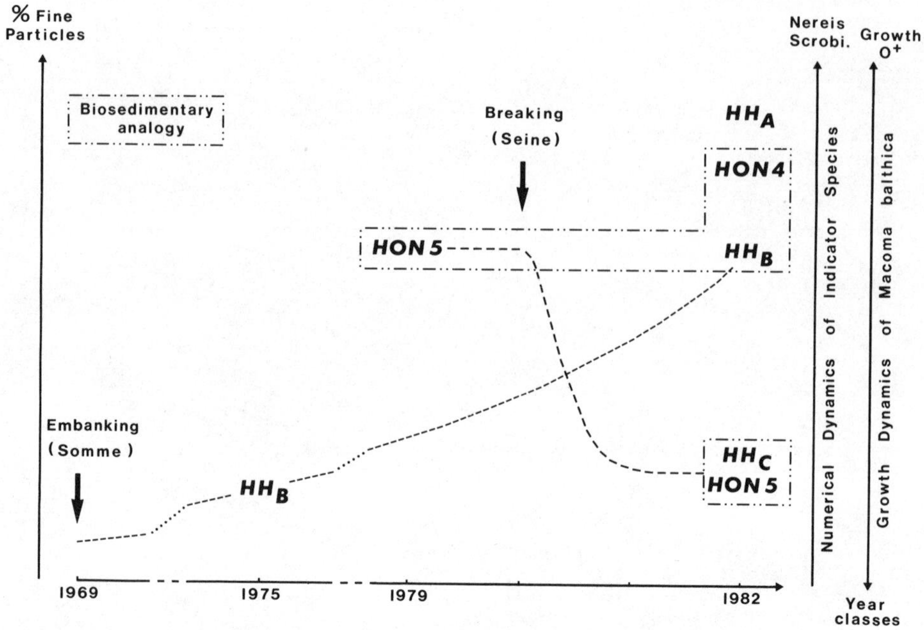


Fig. 15 Diagrammatic representation of biosedimentary evolution in Somme and Seine estuaries

In the last group, it is interesting to mention that the sediment perturbation linked with harbour developments can be understood through a study of Macoma balthica population structure and dynamics in that the fine particle percentage appears to accelerate the growth rate of young individuals and to have a negative influence on the older classes.

## DISCUSSION

The results of Boyden and Little (1973) support the view that changes of faunal composition in the Seine estuary (breaking of the embankment) are the consequence of the loss of fine particles of the sediment. These authors observed that Nephthys cirrosa colonizes the coarse sediments of the Severn estuary and agree with Clark and Haderlie (1960) who suggested that the distribution of this species is basically related to the percentage of various particle sizes.

More recently, in the Ems estuary, Essink (1982) studied the effect of harbour engineering on the nearby sublittoral macrobenthos; his results point out a similar correlation between the benthos evolution and the loss of fine particles resulting from a increase in the local hydrodynamic conditions.

As for the long term biosedimentary evolution (towards the salt-marsh) observed in the Baie de Somme, such a sequence leading to the filling-up of the bay has been described by Le Gall and Larsonneur (1972) in the Baie des Veys; this phenomenon seems typical of the Channel estuaries.

Dupont (1981) pointed out the relations between benthos and intertidal sedimentary phenomena, particularly the important part played by *Pygospio elegans* and *Nereis diversicolor*.

In the Baie de Somme, the bivalve *S. plana* colonizes high shore levels, as found elsewhere by Dankers and Beukema (1981) who point out that this species is typical of high level clay. In contrast, in Honfleur, *S. plana* is found at low shore levels, the situation described by Hughes (1970). Moreover, the disappearace of this species from HON 5, observed since the opening of the embankment, suggests that its distribution is closely related to the sediment granulometry.

Our data on growth rate of *M. balthica* agree closely with those obtained in other European estuaries. We observe three distinct growing periods: growth is maximum during spring, less rapid from July to December, and very slow during winter; but we did not observe the winter growth stop recorded by Warwick and Price (1975) and Beukema and De Bruin (1977). The milder winters explain this phenomenon in our two estuaries where ice is very exceptional on the mud-flats, contrary to the observations of Beukema and De Bruin (1977) in the Wadden Sea. Moreover, Lammens (1967) pointed out that growth did not completely stop in Netherlands during mild winters.

The growth rate of the year class $0^+$ is slightly different from the above, with only two growing periods: growth is maximal from April to December (1 mm month$^{-1}$) and slower in winter (0.55 mm month$^{-1}$). This observation corroborates the results of Chambers and Milne (1975) who pointed out that in Scotland the total growing period is longer among young individuals than among adults.

In the Baie de Somme, the year class $0^+$ grows faster in higher levels of the shore (HH A); the same relation was observed in the Gironde estuary by Bachelet (1980) who explained it by the better stability of the sediment at higher shore levels, in relation to an environment protected from water disturbance by a rocky dam.

In the Seine estuary, the correlation noted between substratum granulometry and the population structure of *M. balthica* agrees with the observations of Reading (1979) that the percentage of young individuals is greater in a fine sediment than in a coarser one. This influence of substratum particle size on the distribution of *M. balthica* can be explained by the relationship between the detritus content of the substratum and its associated micro-organisms upon which *Macoma* feeds (Newell, 1965).

In the Baie de Somme, we obtained contradictory preliminary results: recruitment seemed worse in HH A than in HH C. However, it is important to note that in the Somme estuary, the tidal amplitude reaches 9.80 m and the surveyed stations are distributed between 6,20 m and 8 m. As early as 1940, Rees pointed that the most favourable zones lie below MTL. Ratcliffe et al, (1981), in the Humber estuary, suggested that the spat is scarcer near the schorre; moreover, he noticed that, as in the Somme estuary, the percentage of the year class $1^+$ is higher in upper levels of the shore, due to a passive upshore migration which he demonstrated experimentally. This author pointed out a concomitant upshore scarcity of adults, as observed in the Baie de Somme, and showed experimentally that these are more susceptible to temperature variations.

The conclusions of Kennedy and Mihursky (1971) can also be applied to the upshore situation in the Baie de Somme. They pointed out that the lethal temperature for *M. balthica* lies between 31°C and 35°C according to acclimatization. Their experiments lasted 24 hours, whereas the situation in HH A, where emersion lasts several tidal cycles and temperature can then reach 40°C in summer, is much more extreme. The action of temperature is combined with other factors. Swennen and Ching (1974) pointed out that the big *Macoma* which migrate upward to the surface are infected by a trematode. We observed many individuals crawling on the surface during summer in HH A.

Predation by birds is more prevalent during cold winters when populations of waders are the densest (Caudron et al, 1983) and when *Macoma* begins its upward migration towards the surface (Reading and McGrorty, 1978).

ACKNOWLEDGEMENTS

The authors are grateful to J.J. Beukema for his advice during the study. The manuscript was read by J. Wilson and it is a pleasure to record his help and advice. Thanks are due to the other members of the Groupe d'Etude des Milieux Estuariens et Littoraux for their help in the field and their comments on the text.

Participation to the E.B.S.A. meeting in Dublin was financed by the GEMEL.

Rouen University and Amiens University provided working facilities.

REFERENCES

Bachelet,G. 1980. Growth and recruitment of the Tellinid Bivalve *Macoma balthica* at the southern limit of its geographical distribution, the Gironde estuary (SW France). Marine Biology, 59, 105-117.

Beukema,J.J. 1974. Seasonal change in the biomass of the macrobenthos of a tidal flat area in the Dutch Wadden Sea. Netherlands Journal of Sea Research, 8, 94-107.

Beukema,Biomass and species richness of the macro-benthic animals living on the tidal flats of the Dutch Wadden Sea. Netherlands Journal of Sea Research, 10, 236-261.

Beukema,J.J. and de Bruin, W. 1977. Seasonal changes in dry weight and chemical composition of the soft parts of the tellinid bivalve Macoma balthica in the Dutch Wadden sea. Netherlands Journal of Sea Research, 11, 42-55.

Beukema,J.J. 1981a. Quantitative data on the benthos of the Wadden Sea proper. in: Invertebrates of the Wadden Sea, Marine Zoology, 4, 134-142.

Beukema,J.J. 1981b. The role of the larger invertebrates in the Wadden Sea Ecosystem. in: Invertebrates of the Wadden Sea, Marine Zoology, 4, 211-219.

Bhattacharya,C.G. 1967. A simple method of resolution of a distribution into gaussian components. Biometrics, 23, 115-135.

Boyden,C.R. and Little C.R. 1973 Faunal distribution in soft sediments of the Severn Estuary. Estuarine and Coastal Marine Science, 1, 203-223.

Caudron,E., Ducrotoy, J.-P. and Triplet, P. 1983. Avifaune et macrozoobenthos dans l'estuaire de la Somme: I. Impact de l'Huitrier-Pie Haematopus ostralegus sur les populatins de Coques Cerastoderma edule (Mollusque: bivalve). l'oiseau et la Revue Francaise d'Ornitologie (in press).

Chambers,M.R. and Milne, H. 1975. The production of Macoma balthica (L) in the Ythan Estuary. Estuarine and Coastal Marine Science, 4, 443-445.

Clark,R.B. and Haderlie, E.C. 1960. The distribution of Nephthys cirrosa and Nephthys hombergii on the south western coasts of England and Wales. Journal of Animal Ecology, 29, 117-147.

Dankers,N. and Beukema, J.J. 1981 Distributional patterns of macrozoobenthos species in relation to some environmental factors. in: Invertebrates of the Wadden Sea, Marine Zoology, 4, 69-103.

Desprez,M. 1981. Etude du macrozoobenthos intertidal de l'Estuaire de la Seine. These de Troisieme Cycle, Universite de Rouen: 186 pp.

Dupont,J.-P. 1975. Presence de biolithosores a Pygospio elegans (C) (Polychetes sedentaires) dans la Baie de Somme 80. Compte-Rendu de l'Academie des Sciences, Paris, 280, D, 1773-1776.

Dupont,J.-P. 1981. Relations entre bios et phenomenes sedimentaires intertidaux: le modele de la Baie de Somme. These de Troisieme Cycle, Universite de Rouen: 312 pp.

Essink,K. 1982. The effect of harbour-engineering on nearly sublittoral macrobenthic fauna in the Ems estuary. In: Fluctuation et succession dans les ecosystemes marins, Cabioch, L., Glemarec, M. and Samain, J.F. (Eds.) (in press).

Hughes,R.N. 1970. Population dynamics of the bivalve *Scrobicularia plana* (Da Costa) on an intertidal mud-flat in North Wales. *Journal of Animal Ecology*, 39, 333-356.

Kennedy,V.S. and Mihursky, J.A. 1971. Upper temperature tolerances of some estuarine bivalves. *Chesapeake Science*, 12, 193-204.

Lavoie,R., Tremblay, J.-L. and Filteau, G. 1969. Age et croissance de *Macoma balthica* (L.) a Cacouna-est dans l'estuaire du Saint-Laurent. *Naturaliste canadien* 95: 887-895.

Lammens,J.J. 1967. Growth and reproduction in a tidal flat population of *Macoma balthica* (L.). *Netherlands Journal of Sea Research*, 3, 315-382.

Larsonneur,C. 1977. Classification des sediments superficiels de la Manche *Journal de Recherche Oceanographique*. 2, 33-39.

Le Gall,J. and Larsonneur, C. 1972. Sequence et environnements sedimentaires dans la Baie des Veys (Manche). *Revue de Geographie Physique et de Geologie* (2), 14.2, 189-204.

Newell,R.C. 1965. The role of detritus in the nutrition of two marine deposit feeders, the prosobranch *Hydrobia ulvae* and the bivalve *Macoma balthica*. *Proceedings of the Zoological Society of London*, 133, 25-45.

Petersen,C.G.J. 1913. Valuation of the sea II. The animal communities of the sea-bottom and their importance for marine zoogregraphy. *Reports of the Danish Biology Station* 21,1-44.

Ratcliffe,P.J., Jones, N.V. and Walters, N.J. 1981. The survival of *Macoma balthica* (L.) in mobile sediment. In: *Feeding and survival strategies of estuarine organisms*. (Jones, N.V. and Wolff, W.J. (Eds)) Plenum Press, London and New York, 91-107.

Reading, C.J., 1979. Changes in downshore distribution of *Macoma balthica* in relation to shell length. *Estuarine and Coastal Marine Sciences*, *8*,*1-13*.

Reading,C.J. and McGrorty, S. 1978. Seasonal variations in the burying depth of *Macoma balthica* (L.) and its accessibility to wading birds. *Estuarine and Coastal Marine Science*, 6, 135-144.

Simon,S., Desprez, M., Dupont, J.-P., Peltier, J.-C. and Dupont, W. 1981. Distribution du macrozoobenthos intertidal de la Baie de Somme. *Compte-Rendu de l'Academie des Sciences, Paris*, 292, D, 1013-1016.

Swennen,C. and Ching, H.L. 1974. Observations on the trematodes *Parvatrema affinis*, causative agent of crawling tracks of *Macoma balthica*. *Netherlands Journal of Sea Research*, 8, 108-115.

Warwick,R.M. and Price, R. 1975. Macrofauna production in an estuarine mud flat. *Journal of the Marine Biology Association of the United Kingdom*, 55, 1-18.

# ASSESSMENT OF THE DISTRIBUTION AND AVAILABILITY OF ARSENIC AND MERCURY IN ESTUARIES

W.J. Langston

Marine Biological Association of the United Kingdom
Citadel Hill, Plymouth PL1 2PB, England

## ABSTRACT

Studies have been carried out at sites in British estuaries, covering a wide spectrum of conditions and metal concentrations, in order to recognize factors affecting the distribution and biological availability of arsenic and mercury under natural conditions. Water analyses confirm the importance of particulate iron hydroxides and organic material as the predominant binding substrates for As and Hg respectively. Statistical correlations provide evidence that sediment-bound As and Hg are the major source of these elements to the deposit-feeding bivalves _Scrobicularia plana_ and _Macoma balthica_. to specific binding substrates is used to produce simple predictive models of availability.

A simple estuarine classificatory index comparing the impact of different forms of As and Hg contamination is discussed.

## INTRODUCTION

The theoretical capacity of estuaries to disperse wastes has led to the exploitation of tidal waters as disposal media. Provided adequate knowledge of the transport and fate of discharged material is available it is difficult to argue why such an important resource sould not be utilized. Hence, the powers embodied in various acts such as the Control of Pollution Act (1974), allow for limited control relating to the siting, composition, volume and rate of discharge into tidal waters.

In response to EEC directives dealing with the discharge of dangerous substances to the aquatic environment, the United Kingdom has opted in favour of a monitoring approach to ensure these substances do not reach unacceptable levels as a result of their discharge. Recently, therefore, efforts have been concentrated on the development of environmental quality objectives and specifications for estuarine and coastal waters. For some parameters, such as dissolved oxygen and B.O.D., standards can be set in terms of the amount of pollutant present in the water column. However, for many substances including As, Hg and other metals, levels in water rarely attain concentrations shown to cause effects in laboratory assays, except in the immediate vicinity of waste outfalls. In most cases this is the result of concentration of dispersed substances in sediments, and it is from such sinks that benthic organisms may derive substantially elevated tissue burdens.

Assessment of biological and chemical effects requires the recognition of the interactive behaviour of potential pollutants between dissolved and particulate phases over the range of physiochemical variables encountered in estuaries. Where sediments are a likely source/ sink, it is desirable to produce an effective model relating concentrations in the environment to levels in the tissues of benthic organisms, taking into account factors controlling this process. With the exception of a few studies (see for example, Luoma and Bryan, 1978, 1982; Bryan and Uysal, 1978) remarkably little is understood concerning the availability of sediment-bound metals, especially As and Hg, despite the abundance of literature reporting metal levels in estuarine biota. The purpose of this paper is to report and comment on factors controlling the exchanges between water, sediment and bivalves and to provide a simple means of comparing As and Hg contamination in a variety of estuaries.

## MATERIALS AND METHODS

Deposit-feeding bivalves, surface sediment and, where practical, water samples were collected from a number of sites in estuaries exhibiting a wide variety of physical and chemical conditions (Table 1). Numerous species have been proposed as sensitive indicator organisms for pollution studies and the criteria for selection reviewed by Phillips (1980). For comparative purposes the species must be abundant and widespread and, in the present study area, *Scrobicularia plana* (da Costa) and *Macoma balthica* (L.) have substantial advantages (Bryan et al, 1980).

Bivalves were held in clean 50% seawater for one week prior to analysis. Sediment samples were sieved through 100 m nylon mesh and subsamples extracted with concentrated $HNO_3$, IN-HCl, 25% $CH_3COOH$ and 0.1N NaOH. Total organic content of sediments was derived from measurements of weight loss following ignition at 400$^{\circ}$c for six hours. Precise details of these procedures and A.A. analysis of water, sediment and tissue samples have been described peviously (Langston,

Table 1. Arsenic and Mercury in sediments and bivalves.

| No | Estuary | No of Samples | Sediments (μg/g)[a] | | 'normalized' sediments[b] | | *Scrobicularia plana* (μg/g) | | *Macoma balthica* (μg/g) | |
|---|---|---|---|---|---|---|---|---|---|---|
| | | | As | Hg | As/Fe x10$^3$ | Hg/% organics | As | Hg | As | Hg |
| 1 | Clyde | 13 | 13.2±9.3 | 0.60±1.10 | 0.91± 0.38 | 0.17±0.22 | NS | NS | 16.6±7.7 | 1.19±0.32 |
| 2 | Solway | 3 | 5.1±0.5 | 0.03±0 | 0.64± 0.02 | 0.03±0 | NS | NS | 8.5±0.1 | 0.13±0.05 |
| 3 | Wyre | 9 | 6.4±2.4 | 1.36±1.10 | 0.71±0.19 | 0.53±0.27 | 18.9±6.1 | 0.94±0.23 | NS | NS |
| 4 | Mersey | 28 | 44.9±17.2 | 3.11±1.48 | 2.66±0.90 | 0.79±0.28 | 25.2±5.4 | 1.64±0.54 | 27.1±7.3 | 1.8±1.15 |
| 5 | Dee | 6 | 22.2±9.4 | 1.11±0.56 | 1.46±0.25 | 0.37±0.11 | 23.5±6.4 | 0.78±0.44 | 25.2±14.1 | 1.4±0.13 |
| 6 | Loughor | 13 | 15.8±4.8 | 0.14±0.05 | 1.17±0.16 | 0.03±0.01 | 16.8±4.5 | 0.14±0.07 | 12.1±1.1 | 0.15±0.04 |
| 7 | Swansea Bay | 4 | 15.0±3.3 | 0.47±0.09 | 1.38±0.15 | 0.13±0.02 | 20.4±4.8 | 0.68±0.08 | 22.2 | 0.59 |
| 8 | Neath | 4 | 15.2±5.3 | 0.92±0.33 | 1.17±0.18 | 0.18±0.03 | 21.1±12.0 | 0.63±0.11 | NS | NS |
| 9 | Severn and Bristol Channel | 21 | 8.7±1.36 | 0.46±0.11 | 1.12±0.26 | 0.09±0.02 | 19.0± 5.8 | 0.75±0.68 | 22.6±2.4 | 0.54±0.44 |
| 10 | Taw & Torridge | 4 | 14.0±6.3 | 0.44±0.13 | 1.09±0.23 | 0.09±0.01 | 28.2±9.7 | 0.48±0.16 | NS | NS |
| 11 | Gannel | 2 | 173.8±83.7 | 0.08±0.02 | 11.2±0.4 | 0.02±0.01 | 97.6 | 0.42±0.33 | NS | NS |
| 12 | Hayle | 2 | 749.8±282.7 | 0.16±0.13 | 20.8±1.2 | 0.02±0.02 | 99.9±5.4 | 0.13±0.03 | NS | NS |
| 13 | Restronguet Creek | 3 | 1329 ±250.6 | 0.44±0.15 | 27.1±2.5 | 0.05±0.02 | 176.1±27.5 | 0.07±0.06 | NS | NS |
| 14 | Looe | 10 | 11.7±3.4 | 0.21±0.11 | 1.80±0.43 | 0.06±0.05 | 49.1±18.7 | 0.65±0.30 | 45.7 | 0.97 |
| 15 | Tamar | 11 | 93.5±23.9 | 0.83±0.09 | 3.07±0.47 | 0.13±0.03 | 42.6±14.3 | 0.19±0.03 | 33.9 | 0.27 |
| 16 | Teign | 5 | 31.4±26.2 | 0.30±0.24 | 2.38±1.18 | 0.05±0.03 | 27.8±6.0 | 0.28±0.13 | NS | NS |
| 17 | Poole | 7 | 13.1±4.1 | 2.47±0.70 | 0.56±0.17 | 0.48±0.19 | 11.4±2.7 | 1.16±0.78 | 6.6 | 0.74 |
| 18 | Southampton Water | 14 | 12.6±5.7 | 0.98±0.91 | 0.81±0.41 | 0.20±0.23 | 14.5±5.6 | 0.42±0.30 | NS | NS |
| 19 | Medway | 5 | 18.4±1.7 | 1.0±0.3 | 1.0±0.08 | 0.20±0.04 | 21.3±4.2 | 0.54±0.14 | NS | NS |
| 20 | Thames | 11 | 10.5±6.2 | 0.63±0.37 | 0.58±0.27 | 0.25±0.10 | 19.7±2.7 | 0.86±0.29 | 17.6 | 1.47 |
| 21 | Colne & Blackwater | 8 | 12.1±3.9 | 0.36±0.18 | 0.71±0.11 | 0.11±0.02 | 30.5±9.0 | 0.58±0.14 | NS | NS |
| 22 | Stour & Orwell | 6 | 13.0±4.5 | 0.42±0.18 | 0.72±0.16 | 0.12±0.04 | 30.7±17.0 | 0.52±0.39 | 32.8 9.2 | 0.62 0.14 |
| 23 | Humber | 14 | 46.9±18.0 | 0.53±0.15 | 3.0±2.75 | 0.10±0.02 | 27.6 | 0.24 | 22.6±4.8 | 0.47±0.15 |

a Sediments refluxed with $HNO_3$ for 'total' metal
b As/Fe ratio in 1N HCl extracts

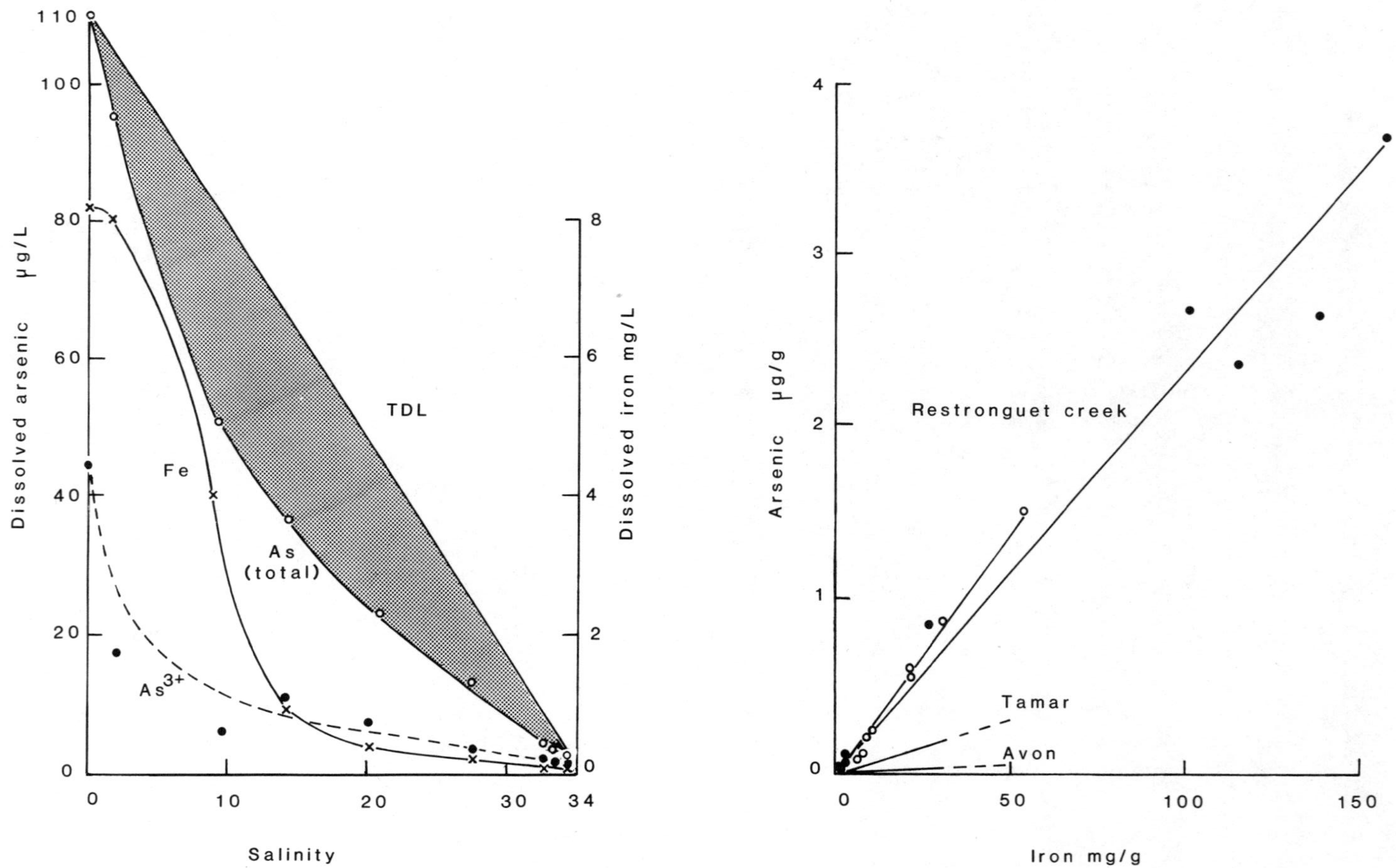


Fig. 1 (A) Behaviour of total dissolved As, $As^{3+}$, and Fe during mixing of Carnon river water and seawater. The shaded area beneath the theoretical dilution line (TDL) shows removal of dissolved As. (B) Correlations between As and Fe in suspended (●) and bottom (○) sediments of Restronguet Creek. Regression lines for Tamar and Avon (Devon) bottom sediments are included for comparison. (From Langston, 1983).

1980, 1982, 1983). All sediment and tissue concentrations reported are calculated on a dry weight basis.

The results for individual sites have been pooled to provide mean values for each estuary in order that broad comparisons of contamination could be made. It is acknowledged that inter-site variation does occur within estuaries although such variation, particularly in the normalised sediment values used to produce the indices of contamination, is usually low. Bryan et al (1980) have also demonstrated the uniformity of metals in sediments and biota over large areas of inter-tidal mudflats.

Correlation and regression analyses were performed on the data to evaluate the significance of water-sediment-organism interrelationships for As and Hg, in the various extractants used. Separate analyses were performed to assess the effects of other variables (Fe, Mn, Ca, organic matter) on these associations.

## RESULTS

### Arsenic

Sources and distribution. The geochemistry of As in estuaries is poorly understood despite its inclusion among "grey-listed" toxic substances. It is found naturally in arsenopyrite (FeAsS) commonly associated with the mineralized zones of south west England. Production of arsenic, essentially as $As_2O_3$, takes place as a by-product of base metal refining. Its uses include alloys, semiconductors and wood preservation, although it is primarily used in a variety of herbicide, algicide and pesticide applications as a result of its toxic qualities (Newland, 1982).

Water analyses implicate run-off from mineralized catchment areas as the most significant input of As into estuaries, although localised effects of metal refining operations can be equally important (Table 2). The Carnon river which drains into Rostronguet Creek contains the most consistently elevated levels of As largely due to the influx of metal-rich, low-pH waters of the County Adit. Removal of dissolved As by coprecipitation with freshly formed Fe(oxy) hydroxides is a significant feature during estuarine mixing (Fig. 1A) and accounts for the strong correlations between As and Fe in suspended ($r^2$ = 0.9301) and surface ($r^2$ = 0.9948) sediments in Restronguet Creek (Fig. 1B). It is calculated that 80-100% of incoming As, amounting to some 9.3 tonnes annually, is retained by creek sediments, which are, therefore, likely to be the major surce of As to benthic organisms (Langston, 1983). A similar removal mechanism is believed responsible for the strong correlations between As and Fe in sediments from many British estuaries (Langston, 1980, 1983). The

Table 2 A comparison of As and Hg concentrations in water samples[a]

| Location | As(μg/l) | Hg(ng/l) |
|---|---|---|
| Plymouth (seawater) | 1.3 | 5 |
| Mersey (estuarine) | 2-8 | 10-30 (91-99) |
| Holes Bay, Poole (estuarine) | 2.1 | 5.53 (662-98) |
| Wyre (estuarine) | 1.3 | 5 (98) |
| Southampton Water (estuarine) | 2.3 | 12[b] |
| Tamar (estuarine) | 2-9.5 | 10-40 (72-92) |
| Humber (estuarine) | 45-13,000 | 15-370 |
| Thames (estuarine) | 1.8 | 3-18 (72)[c] |
| Carnon River (freshwater) | 29-280 | 30 |
| Restronguet Creek (estuarine) | 3-95 | --- |
| Hayle (freshwater) | 15-19 | 10 |
| Hayle (estuarine) | 4.6 | 10 |
| Gannel (freshwater) | 3.0 | 10 |
| Gannel (estuarine) | 1.6 | 10 |
| Severn/Bristol Channel (estuarine) | 2.2 | 10-50[d] |
| Teign (estuarine) | 1.4 | 10 |
| Taw/Torridge (estuarine) | 1.5 | 10 |
| Looe | 1.6 | 10 |

a) Data for soluble metals except for Humber samples and Hg in Southampton water which include dissolved and particulate fractions. Figures in parentheses are percentages of Hg associated with particulates.
b) Data from Burton and Leatherland (1971)
c) Data from Nelson (1981)
d) Data from Gardner and Riley

regression lines for Tamar and Avon sediments are compared with those of Restronguet Creek in Figure 1B.

Concentrations of metals in sediments remain stable for long periods and are therefore good integrators of contamination in comparison to the rapidly fluctuating conditions in estuarine waters. A considerable range of As concentrations is evident in the present study (Table 1). The lowest values, 5-10 μg/g, occur in the Solway and Wyre, while sediments from the majority of estuaries contain 10-20 μg/g. Some enrichment is evident in Humber and Mersey samples, though of much less significance than in the estuaries of south west England. Highest concentrations of As in surface sediments, 2,500 μg/g, were found in the region of freshly precipitated particles in Restronguet Creek.

Availability: Reliable data for As in bivalves is scarce. Whole animal concentrations are generally less than 10 μg/g (Eisler, 1981),

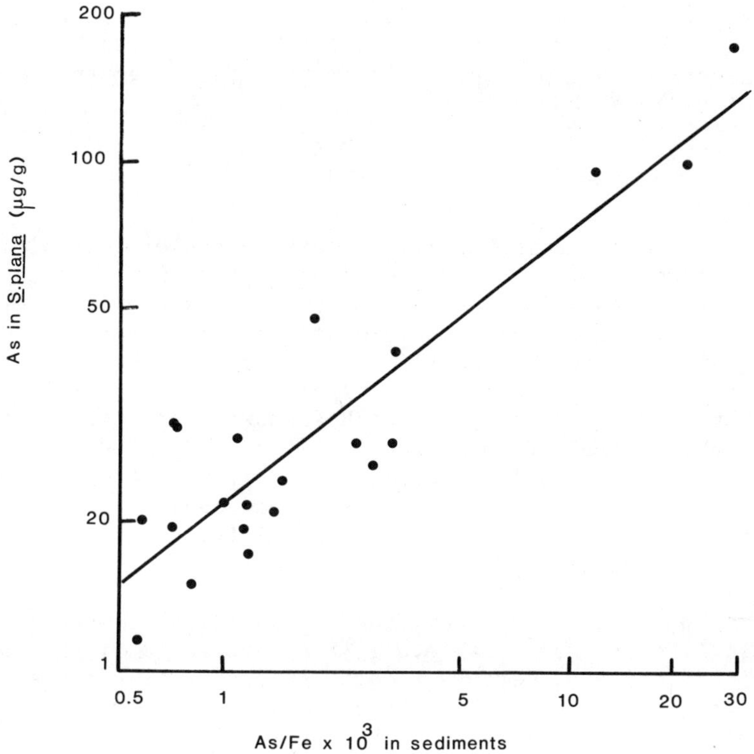


Fig. 2 Relationship between As in S. plana and As/Fe ratio in n-HC1 sediment extracts (logaritmic scale). o Equation for line is:
As in S. plana = 5.3352 As/Fe in sediments + 17.6757, n = 21, r = 0.96
$\log_{10}$ As in S. plana = 0.5494 $\log_{10}$ As/Fe in sediments + 1.3317, n = 21, r = 0.90 for transformed data

although individual tissues of some species contain higher levels. Concentrations of As in S. plana range from 11.4 - 176 µg/g (Table 1) and generally reflect sediment values rather than dissolved As. Accumulation is notably greatest in Restronguet Creek, Hayle and Gannel animals. Correlations between As in sediments (HC1 extracts) and tissues were improved by the inclusion of extractable Fe as a variable (Figure 2, r = 0.96, $P < 0.001$) in close agreement with earlier results for individual sites (Langston, 1980). As sediment concentrations normalized to Fe in IN-HC1 extracts therefore provide the best predictable information of availability.

The results for M. balthica, though limited, are remarkably similar to those for S. plana (Fig. 4a). As concentrations in M. balthica correlate most significantly with As/Fe ratios and the slope and intercept (0.5714 and 1.2468 respectively for transformed regression data) are comparable to values for *S. plana* (see Legend Fig. 2). Fe is therefore an important environmental factor governing availability of sediment-bound As.

## Mercury

Sources and distribution: There are no exploitable mercury ores in the watersheds of the estuaries studied and the major inputs are therefore anthropogenic. Mercury is used by the chemical, electrical, pharmaceutical and dental industries and is discharged either directly into waterways or transported for disposal in the form of sewage sludge and industrial spoils. In total, UK industry is supplied with some 700 tonnes Hg annually of which approximately two thirds is consumed by chlor-alkali plants (D.O.E., 1976). Despite reductions in the discharge of Hg, and improvements in effluent treatment, leaking industrial waste deposits still persist and sediments contaminated from previously uncontrolled discharges will continue as a source of metal for many years.

Water analyses demonstrate the partitioning of mercury in favour of particulate material (Table 2). The importance of suspended solids in determining mercury budgets in estuaries has been demonstrated in a number of studies (see for example Nelson, 1981; Airey and Jones, 1982; Rae and Aston, 1982).

Figures for Hg in water at best only serve to pick out exceptional 'hot spots' and the dominance of particulate Hg makes sediment measurement a more useful yardstick for monitoring purposes.

The strong affinity of Hg, particularly for organic ligands results in good correlations between Hg and organic material in many British estuaries (Langston, 1982). The distribution of Hg in sediments may therefore be a function of organic content in relation to the nature and siting of the input.

Mean Hg values in surface sediments range over two orders of magnitude from 0.03 μg/g in the relatively unpolluted Solway to 3.1 μg/g in the Mersey estuary (Table 1), where concentrations at individual sites attain a maximum of 6.2 μg/g. Discharges into the Mersey occur from a large number of domestic and industrial sources including the chlor-alkali industry. In addition, dumping of dredged material and sewage sludge may be important in the redistribution of Hg in this area (Airey and Jones, 1982). Inputs from the chlor alkali industry are probably the major source of Hg into sediments of the Wyre Estuary (Rae and Aston, 1982) which contain up to 3.71 μg/g. Comparable levels of contamination are also found in sediments from Holes Bay, Poole, where potential sources include chemical works, light industry, docks, marinas and a major sewage treatment plant.

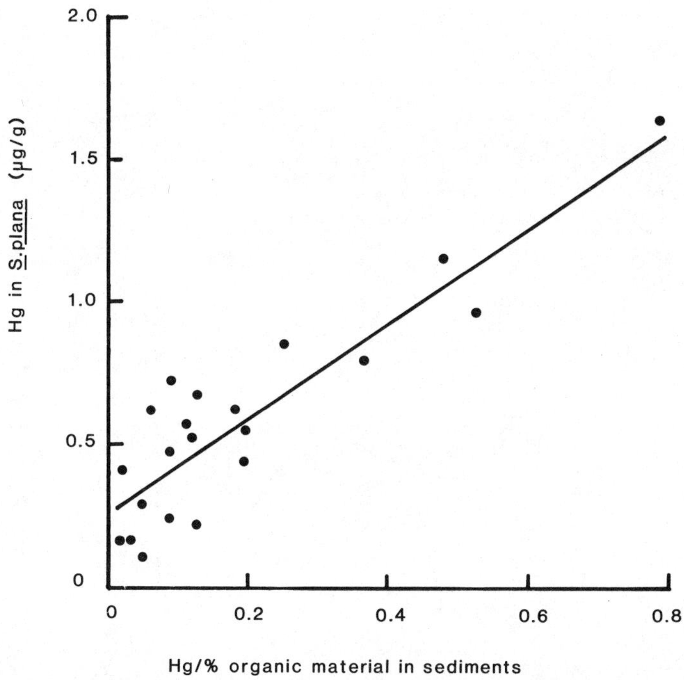


Fig. 3 Relationship between Hg in S. *plana* and Hg/% organic material in sediments. Equation for line is:
Hg in *S. Plana* = 1.6631 Hg/% organics in sediment + 0.2602, n=21, r=0.88

Availability: Bivalves are frequently used as indicators of mercury pollution and the mussel-watch surveys are useful in providing a general picture of the spread of contaminants (see for example De Wolf, 1975; Goldberg et al., 1978, Murray, 1982). Direct proportionality between environmental levels and tissue residues is frequently assumed however, and the influence of modifying factors largely ignored. Baseline Hg concentrations in *S. plana* and *M. balthica*, as in many other molluscs (Eisler, 1981) are of the order of 0.1 μ g/g and are exceeded by a factor of ten in animals from the Mersey, Wyre and Poole. Highest concentrations of 2.5 and 2.4 μ g/g Hg were present in *S. plana* at specific sites in the Mersey and Poole Harbour, and 4 μ g/g in *M. balthica* from an upstream site in the Mersey.

Hg burdens in *S. plana* and *M. balthica* can largely be attributed to sediment concentrations (r = 0.81, 0.57) although some notable anomalies suggest the influence of other factors on accumulation. The organic content of sediments is the most influential of the variables studied, improving correlations for *S. plana* (r = 0.88) and *M. balthica* (r = 0.74). The relationship between Hg in *M. balthica* has

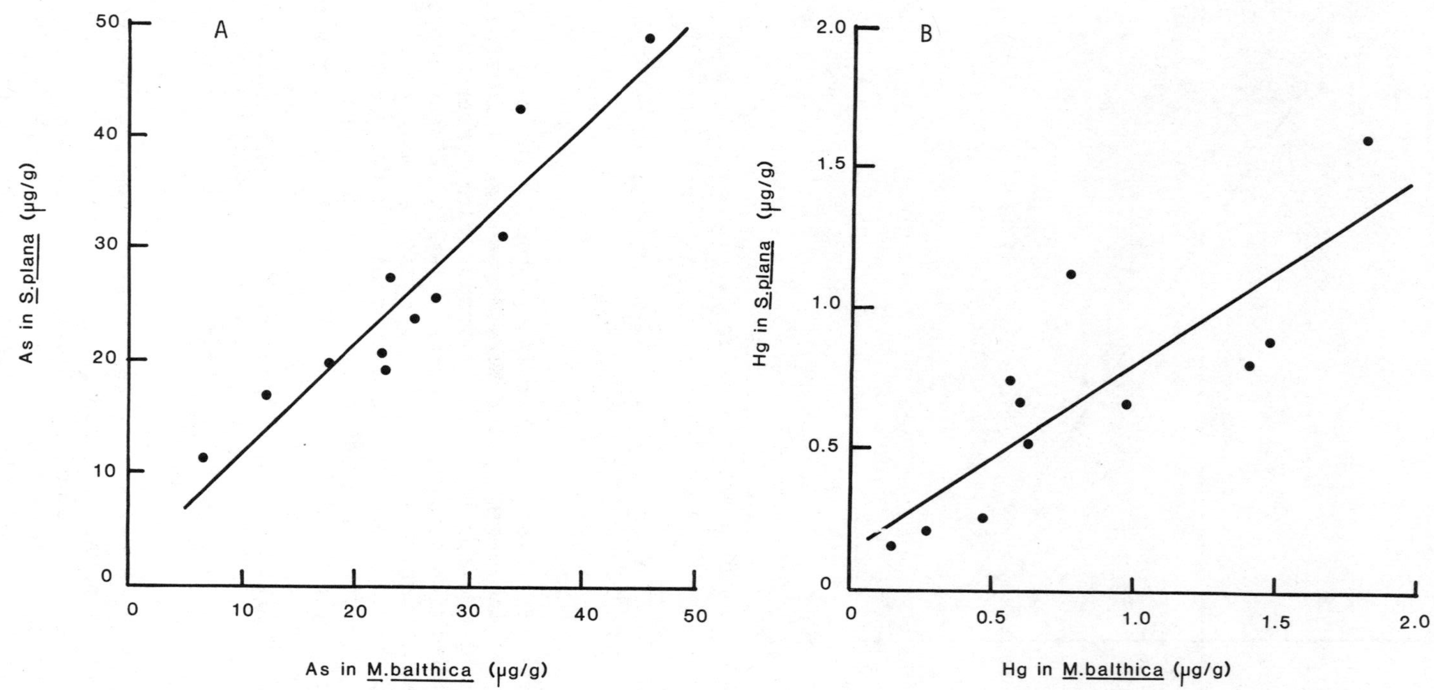


Fig. 4 (A) Relationship between As in M.balthica and S.plana. Equation for line is : As in S.plana = 0.9781 As in M.balthica + 2.1345, n=11, r=0.93.
(B) Relationship between Hg in M.balthica and S.plana. Equation for line is : Hg in S.plana = 0.6697 Hg in M.balthica +0.1427, n=11, r=0.80.

values of 1,7899 and 0.4181 for slope and intercept respectively. Availability of Hg in both M. balthica and S. plana is therefore modified by the organic content of sediments, and proportionality between concentrations in the two species is, not surprisingly, evident. However, the value of 0.6697 for the slope of this line (Fig. 4b) suggests M. balthica may be the better accumulator of Hg.

## DISCUSSION

The complexity of variables influencing the processes of distribution and availability of pollutants in estuaries, severely handicaps the production of truly comprehensive models. Faced with such complexity it is essential firstly to identify the major variables which determine the fate of specific compounds.

Difficulties in obtaining representative water samples, together with problems associated with analysis of low levels have focused attention on benthic organisms as integrators of environmental and, especially, sediment contamination. Nevertheless understanding the water-sediment interactions of potential contaminants is important before evaluations of availability can be made. In this study the role of sediments as the major respository of estuarine inputs of As and Hg has been assessed. Freshly precipitated Fe(oxy) hydroxide is considered to be the most effective binding substrate for As and is therefore most likely to affect its distribution in estuarine sediments. Although Fe may dominate the partitioning of As, some association with other phases such as organics cannot be discounted, since reactivity of Fe and organics are frequently closely connected in estuaries. In contrast, Hg distribution in oxidised surface sediments is most strongly determined by interactions with organic matter, an association encountered in a variety of estuaries (see for example Crecelius et al, 1975; Lindberg et al, 1975; Langston, 1982; Rae and Aston 1982.

S. plana and M. balthica are among the dominant members of the benthic infauna in the majority of estuaries studied. Though not commercially exploited, they are important food sources for a variety of predators and scavengers and, as deposit feeders, play a significat role in the remobilisation of contaminants in sediments. Sediment concentrations of As and Hg are the most important of the parameters studied correlating with metal burdens in both species. Strictly accurate predictions of availability could not, however, be based on sediment concentrations alone.

The substrates which determined partitioning of As and Hg in sediments, Fe (oxy) hydroxides and organics respectively, also play a significant role in modifying availability. It may seem obvious that a metal should partition predominantly to the substrate with the highest bonding strength and hence lowest availability, yet this may not be the case for all metals (Louma and Bryan, 1982).

Simple models based on partitioning of As and Hg in sediments can be constructed, which statistically correlate with availability in deposit feeders. 1N-HC1 extracts provide the best assessment of As availability and are useful generally in identifying the extractable phase of most metals (Luoma and Bryan, 1982). However, a much stronger attack (concentrated nitric acid) is needed to quantify Hg in sediments, which is resistent to extraction with 1N-HC1 due to the strength of the organic-Hg bonds (Langston, 1982). The strength of these bonds explains the inverse relationship observed between Hg availability and organic content in sediments.

Other parameters must inevitably be controlling availability in the wide range of conditions present in these estuaries. However the simple statistical comparisons made in this paper suggest the most significant factors have been identified.

Normalisation techniques have been employed frequently to reveal metal enhancement in sediments (Calvert and Price, 1977; Wright et al, 1977; Kristensen, 1982) though little attention has been paid to defining these values in terms of biological availability. The demonstration in this paper that As and Hg in tissues is quantifiably related to normalised sediment concentrations enables inter-estuarine comparisons of availability and contamination to be made. Predicting tissue levels is also possible in the absence of suitable indicators. The specific linear relationships between Hg and organic matter, or As and Fe, observed for many estuarine sediments allows the characterisation of availability on the basis of a single values, or the slope of the regression line.

An index of contamination based on these normalised sediment values is depicted for As in Figure 5 and Hg in Figure 6. Estuaries most contaminated with As lie within the mineralized areas of south-west England, the only region in Great Britain to have yielded As in quantity. Between four and eight thousand tons were produced annually from 1870-1902, mainly in the Camborne, Redruth, St Day area of Cornwall (40%) and Callington and Tavistock in Devon (58%). With the exception of South Crofty at Camborne and Mount Wellington and Wheal Jane in the Carnon Valley, the majority of mines have long since ceased production (Dines, 1956). The resulting affect on its availability is shown in Figure 5, which highlights the high concentrations found in Restrongest Creek and the Hayle and Gannel estuaries. The Tamar, the scene of half the world's As production a century ago, stll contains elevated levels as a result of As-rich sediments together with inputs from mine spoils and adit drainage (Aston et al, 1975, Langston, 1983). As was mined to a comparatively minor extent in the Teign valley compared with more westerly regions.

The appearance of the Humber and Mersey alongside the mineralised estuaries in this index indicates some degree of contamination as a result of the large quantities of urban and industrial wastes

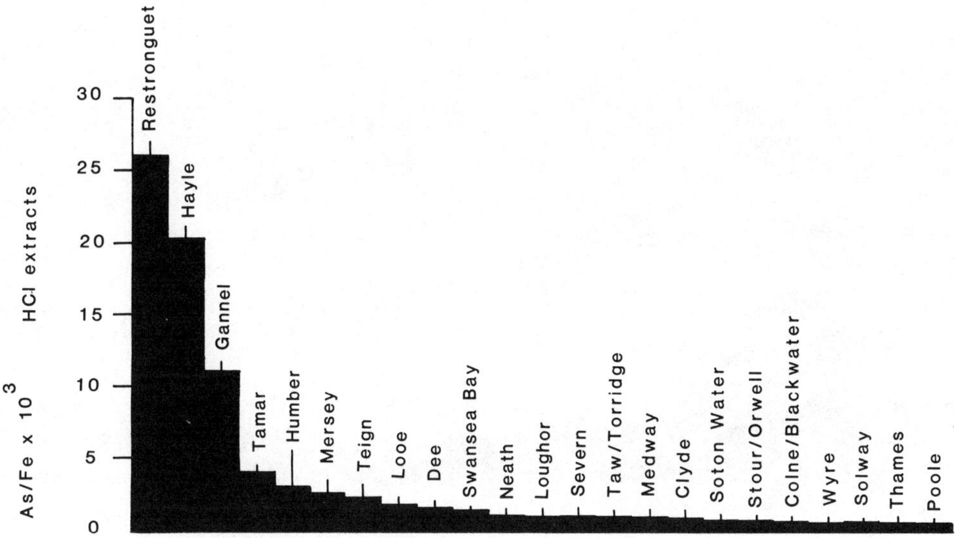


Fig. 5 Index of As availability based on Jn-HCl extractable sediment concentrations 'normalised' to Fe. Standard deviations indicated.

received. Although mean As/Fe ratios are an order of magnitude lower than Restronguet Creek, values up to 11.1 were found near a metal refinery outfall in the Humber, and may be locally significant in terms of uptake and effects on biota (Goulder et al., 1979; Langston, 1980).

Availability of As is comparatively low in other estuaries and the position of Poole and the Thames estuary in the index suggest sewage inputs, which may be important in both areas, is of little significance in terms of As contamination.

Mineralization and mining in south west England has produced little effect on Hg availability, and the estuaries in this area are among the least contaminated of those studied (Fig. 6). Not surprisingly, abundance and availability in sediments is greatest in those estuaries receiving waste from the largest users of Hg, the chlor-alkali plants on the Mersey and Wyre. However, sediments from Poole, with its inputs from the chemical industry and sewage, are only slightly lower in terms of available metal.

The watershed of the Dee incorporates fairly large urban and industrial developments as likely sources of Hg. Nevertheless, the status of this estuary in the present index may be partly due to the influence of inputs from the Mersey and Liverpool Bay. Evidence for this hypothesis is derived from results of sediment and tissue analyses which indicate an atypical decrease in Hg concentrations with distance upstream in the Dee.

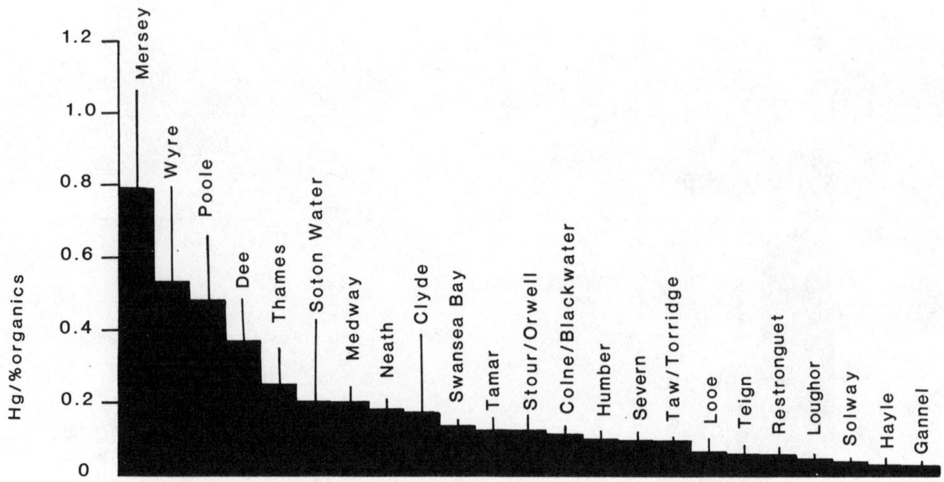


Fig. 6 Index of Hg availability based on sediment concentrations 'normalised' to % organic material. Standard deviations indicated.

The major input of mercury into the Thames estuary is sewage effluent (1.7 Kg Hg/day) and sewage sludge dumping (3.9 Kg Hg/day) (Nelson, 1981). It is therefore interesting that availability (Hg/% organics) in Thames surface sediments (0.25) is remarkably similar to the calculated value (0.23) for sewage sludge dumped in the estuary (Norton et al, 1981). Sewage waste may therefore have contributed significantly to the elevated Hg levels found in Thames bivalves although continued dumping of similar material is unlikely to cause further increases.

Availability of Hg in the remaining estuaries is reasonably consistent with existing data of known inputs, although the relatively uncontaminated status of the Humber and Severn/Bristol Channel is, perhaps, unexpected.

In summary, simple models are presented which are capable of adequately describing availability of As and Hg to deposit feeding bivalves. Tissue concentrations of these potential pollutants can generally be predicted in organisms from sediment concentrations normalised to the amount of specific binding substrate. Consequently for a variety of estuaries differing widely in physiochemical conditions. Using a similar approach it sould be possible to predict the impact of future discharges and to establish suitable criteria to prevent the accumulation of substantially elevated levels in the biota.

ACKNOWLEDGEMENT

This work was financially supported by the Department of the Environment under contract DGR 480/51.

REFERENCES

Airey, D. and Jones, P.D. 1982. Mercury in the River Mersey, its estuary and tributaries during 1973 and 1974. Water Research 16, 565-577.

Aston, S.R., Thornton, I., Webb, J.S.,Milford, B.L. and Purves, J.B. 1975. Arsenic in stream sediments and waters of south west England. Science of the Total Environment 4, 347-358.

Bryan, G.W. and Uysal, H. 1978.Heavy metals in the burrowing bivalve Scrobicularia plana from the Tamar estuary in relation to environmental levels. Journal of the Marine Biological Association of the United Kingdom, 58, 89-108.

Bryan, G.W., Langston, W.J. and Hummerstone, L.G.1980. The uses of biological indicators of heavy metal contamination in estuaries. Occasional Publications. Marine Biological Association of the United Kingdom, No. 1, 73 pp.

Burton, J.D. and Leatherland, T.M. 1971. Mercuryin a coastal marine environment. Nature, 231, 440-442.

Calvert, S.E. and Price, N.B. 1977. Geochemicalvariation in ferromanganese nodules and associated sediments from the Pacific Ocean. Marine Chemistry 5, 43-74.

Crecelius, E.A., Bothner, M.H. and Carpenter, R, 1975.Geochemistries of arsenic, antimony, mercury, and related elements in sediments of Puget Sound. Environmental Science and Technology, 6, 274-278.

Department of the Environment.1976. Environmental mercury and man. H.M.S.O., London.

Dines, H.G. 1956.The metalliferous mining region of south-west England, vol. 2. H.M.S.O., London, 795 pp.

Eisler, R. 1981.Trace metal concentrations in marine organisms. Pergamon Press, 687 pp.

Gardner, D. and Riley, J.P. 1973.The distribution of dissolved mercury in the Bristol Channel and Severn estuary. Estuarine and Coastal Marine Science 1, 191-192.

Goldberg, E.D.,Bowen, V.T., Farrington, J.W., Harvey, G., Martin, J.H., Parker, P.L., Risebrough, R.W., Robertson, W., Schneider, E. and Gamble, E. 1978. The mussel watch. Environmental Conservation 5, 101-125.

Goulder, R., Blanchard, A.S., Metcalf, P.J. andWright, B. 1979. Inhibition of estuarine bacteria by metal refinery effluent. Marine Pollution Bulletin 10, 170-173.

Kristensen, P. 1982.Time-dependent variation of mercury in a stream sediment and the effect upon mercury content in Gammarus pulex (L.). Water Research 16, 759-764.

Langston, W.J. 1980.Arsenic in U.K. estuarine sediments and its availability to benthic organisms. Journal of the Marine Biological Association of the United Kingdom 60, 869-881.

Langston, W.J. 1982.The distribution of mercury in British estuarine sediments and its availability to deposit feeding bivalves. Journal of the Marine Biological Association of the United Kingdom 62, 667-684.

Langston, W.J. 1983.The behaviour of arsenic in United Kingdom estuaries. Canadian Journal of Fisheries and Aquatic Sciences (in the press).

Lindberg, S.E., Andren, A.W. and Harriss, R. 1975.Geochemistry of mercury in the estuarine environment. In Estuarine Research, vol. 1 (Cronin, L.E. (Ed)) Academic Press. pp. 64-108.

Luoma, S.N. and Bryan, G.W. 1978. Factors controlling the availability of sediment-bound lead to the estuarine bivalve Scrobicularia plana. Journal of the Marine Biological Association of the United Kingdom 58, 793-802.

Luoma, S.N. and Bryan, G.W. 1982. A statistical study of environmental factors controlling concentrations of heavy metals in the burrowing bivalve Scrobicularia plana and the polychaete Nereis diversicolor. Estuarine, coastal and shelf science 15, 95-108.

Murray, A.J. 1982.Trace metals and organochlorine pesticide and PCB residues in mussels from England and Wales, 1978. Chemistry in Ecology 1, 33-45.

Nelson, L.A. 1981.Mercury in the Thames estuary. Environmental Technology Letters 2, 225-232.

Newland, L.W. 1982. Arsenic, beryllium, selenium and vanadium. In: The handbook of environmental chemistry. Vol. 3, Part B. Springer-Verlag pp. 27-67.

Norton, M.G.,Eagle, R.A., Nunny, R.S., Rolfe, M.S., Hardiman, P.A. and Hampson, B.L. 1981. The field assessment of effects of dumping wastes at sea: 8 Sewage sludge dumping in the outer Thames estuary. Fisheries Research Technical report, M.A.F.F. Directorate of Fisheries Research, Lowestoft (62) 62 pp.

Phillips, D.J.H. 1980. Quantitative Aquatic Biological Indicators. London: Applied Science Publishers. 488 pp.

Rae, J.E. and Aston, S.R. 1982. The role of suspended solids in the estuarine geochemistry of mercury. Water Research 16, 649-654.

Wolfe, P. De. 1975. Mercury content of mussels from West European coasts. Marine Pollution Bulletin 6, 61-63.

Wright, D., Soner, J.H. and Chester, R. 1977. Some preliminary results of trace metal studies in the Loughor estuary and Burry Inlet. In Problems of a small estuary (Nelson-Smith, A. and Bridges, E.M. (Eds)) Quadrant Press, Swansea. pp 4: 1/1 - 4: 1/21.

# THE APPLICATION OF TWO SIMPLE INDICES TO IRISH ESTUARY POLLUTION STATUS

D.W. Jeffrey, J.G. Wilson, C.R. Harris and D.L. Tomlinson

Environmental Sciences Unit
Trinity College, Dublin

## ABSTRACT

Two complementary indices of estuarine quality have been developed as a management tool and to facilitate intercomparison of diverse estuarine types. The indices used here have been devised as relatively quick and inexpensive and taking into account both the biological and the environmental information from a systematic survey of the intertidal regions of an estuary.

The Biological Quality Index (BQI) is based on the proportions of the estuary classified abiotic, opportunistic or stable. The Pollution Load Index (PLI) is obtained from analysis of the intertidal sediments for pollutants which are then scored according to baseline (i.e. unpolluted) and threshold (i.e. at which adverse biological effects occur) values.

The BQI and PLI were then tested in seven Irish estuaries which represented the range of sizes and industrialisation present in Ireland. The results for BQI ranged from 0.0 (completely abiotic) to 9.94 (largely unpolluted), and for PLI from 0.0 to 2.56, although individual site PLI values ranged up to 10.0.

Modification of the indices and their wider application to other European estuaries are discussed.

## INTRODUCTION

The two indices to be described are an important part of a method for describing and intercomparing the quality of estuarine environments. In order to evaluate the method and the indices it is necessary to state the objectives in mind when they were developed, notably communication between scientists and other interested parties including administrators and the public leading eventually to their application as a management tool.

These objectives were a major constraint not encountered in the monitoring of individual estuaries, and are outlined more fully below.

a) The method should be applicable throughout western Europe. It is hoped that the specific details of the procedure are applicable at least within these territorial limits, whilst the general principles outlined may be used universally.

This synoptic dimension alone potentially restricts use of biological indicator procedures.

b) Environmental quality should be determinable at any season of the year, certainly whenever fieldwork is possible.

c) Procedures and methods should be within the technical capability of a wide variety of agencies.

This constraint seriously restricts the assimilation of the more complex research methods requiring specialist knowledge, techniques or equipment such as the sampling and analysis of interstitial fluids from sediments (e.g. Nurnberg, 1980) or the various uses of specific biological indicators. The indexing procedure is however sufficiently flexible to assimilate data from such techniques if they are developed for widespread use.

d) Procedure should be rapid and economical.

Given the urgency with which answers are required and the financial constraints which obtain, it was decided to devise the most rapid and economical method consistant with scientific accuracy. It seems particularly important to avoid wasteful oversampling or over-analysis. A problem that arises from economics of this kind is the comparison of sites with different kinds of data base.

e) The need for a suitable indexing procedure was strongly indicated by the desirability of the public's comprehension of statements of estuarine quality as well as facilitating the interpretation of the data by both environmental specialists and non-specialised administrators. While it is understood that any index represents a simplification of the situation, a good index will provide, at the very least, a direction for further or more intensive investigation.

## PROPERTIES OF INDICES

The main purpose of any index is communication, that is to condense, integrate and interpret a mass of data into a easily understood form. They are widely used in many fields, for example various indices used to chart the state of the stock markets, and other indices are in use as a indication of air pollution (Thomas, 1975).

To be of use, the index must reflect the values and trends of the data in general. Most indices do this by selecting, from the mass of data, some element or elements within it such that change within the total mass of data is accompanied by a corresponding change in the element(s) chosen. Such a selection procedure is obviously more accurate when there is information available as to the comparative reactions of the data mass and the chosen elements. Furthermore, the chosen element(s) should integrate the data so that as little as possible of the total information available is lost. Obviously the more elements chosen, the better the integration, but a balance has to be struck between the absolute accuracy of the final index and its comprehensibility. Operational experience leads to better understanding of the behaviour of an index, and confidence in its use as an interpretative tool.

The final form of the index is usually numerical, since a numerical value gives a precision and objectivity to the index which would be lacking in a verbal description, and this is greatly preferred by non-specialists in the field who nevertheless wish to know what is happening.

The various environmental quality indices that exist, whether air pollution, fresh water pollution or marine pollution, are based on the concept of an unpolluted baseline, that is for every variable, there are values representative of those which would occur in an unpolluted situation. These variables may be chemical as in many air quality indices (Babcock & Nagda, 1975) or biological as in many fresh water or marine pollution indices. The latter can be further divided into two main categories:

1) population responses of species, numbers or diversity indices (e.g. Cook, 1976; Gray, 1979)

2) individual physiological response indices (e.g. Bayne et al, 1979).

The two indices to be described, the Biological Quality Index (BQI) and Pollution Load Index (PLI) are also based on comparisons with the concept of unpolluted biological and chemical baselines.

## DATA COLLECTION

In order to comprehend the indices it is first necessary to briefly describe the standard method of data collection. This procedure hinges on the selection of the sample sites, which utilises the stratified random positioning of the sites. The advantages of such an approach have been discussed by Hiscock (1979) and can be stated briefly as avoiding over-sampling, and thus the over-commitment of time and money, while ensuring that all conditions are adequately represented.

In the stratified random approach the sample sites are placed randomly within a homogeneous area, which for these indices correspond to the substrate or sediment type with their characteristic associated biological communities. Two points may be noted here in relation to these facies or zones. Firstly, evidence of heterogeneity, for example different species lists from different sites within one zone, indicates the estuary has been incorrectly divided and further sampling, or complete re-sampling is needed. Secondly, and for the above reason, it has often been found advantageous to classify areas adjacent to known or suspected pollution sources as separate zones and sample them as such.

In practical terms this means that two days sampling, say 20 sites, may be divided between e.g. either 4 zones (5 sites each) in the estuary or 5 zones (4 sites each) to include a suspected polluted zone. The BQI and PLI are then calculated from standardised descriptions of the chosen sites.

## BIOLOGICAL QUALITY INDEX (BQI)

The BQI describes the biological health of the estuary as a function of the macrobiota present. In a unpolluted estuary, although the range of species may be limited, there will be reasonably abundant and stable populations of bivalves such as _Scrobicularia plana_ and _Macoma balthica_ and the polychaete _Nereis diversicolor_ in the mud and of mussels and fucoids on solid substrates. Under severe pollution all macrobiotic life disappears and this condition is easily recognised and has often been described (Pearson and Rosenberg, 1978). The problem lies in the detection of the pollution status in between that of stable communities and abiotic environments and a number of approaches have been tried in the marine environment. These include the use of the log-normal distribution (Gray, 1979) and the use of diversity indices such as the Shannon-Weiner (Gray, 1976) or indicator species (Eagle and Rees, 1973), and similar approaches which worked well in fresh water (Cook, 1976). The indicator species approach has been refined by various authors including Leppakowski (1975) and the literature review by Pearson and Rosenberg (1978) lists these opportunistic or transgressive species. These are mainly small, short-lived species which have a high reproductive capacity - in other words "r" strategists (Gray, 1979).

In any physically-controlled environment such as a estuary, the tendency will be for the biota to be dominated by "r" strategists (Sanders, 1968; Gray, 1979), and in fact several of the opportunistic species listed in Pearson and Rosenberg (1978) are found in unpolluted estuaries. The problem is compounded in high-energy, totally physically-controlled estuaries. Nevertheless, it was felt that this was a valuable and easily understood method of defining areas, which are neither stable nor abiotic, by their domination by small, short-lived opportunistic species. A similar approach has been evolved by Leppakowski (1975, 1977) in which he has defined areas of primary and secondary maxima and minima corresponding to the progressive and regressive species of the 1st and 2nd orders with distance from the pollution source. Leppakowski (1975, 1977) noted that two density peaks could be observed in studies of animal communities round a pollution source, the secondary maximum closer to the polluted area, which comprised species showing marked signs of opportunism and the primary maximum at some distance from the pollution source, at wich biomass, if not numbers of species or individuals, was at its maximum. The Benthic Pollution Index was then derived from the relative areas which could be assigned to the different categories. Accordingly, it was decided that each zone within the estuary would be assigned to one of the following categories: abiotic (A), that is no macrobiotic life; opportunistic (B), that is dominated by small, short-lived, opportunistic species; or stable (C) with either considerable species diversity or evidence of stable long-term conditions (e.g. several year classes present). The proportional areas for each assigned category are then determined for each zone such that

$$A + B + C = 1.0.$$

The BQI is then calculated from the formula

$$BQI = \text{antilog}_{10}(C - A),$$

where C = proportional area stable and A = proportional area abiotic. The estuary BQI is obtained by the addition of each zone BQI multiplied by the proportional area of their respective zones.

For example an estuary with a total area of 10 km$^2$ may have a BQI as shown in Table 1.

Here

$$\text{estuary BQI} = \text{antilog}_{10}(0.70 - 0.05) = 4.47.$$

Some worked examples from Irish estuaries are included later.

Table 1 Zonation of a hypothetical estuary.

| Zone | Area ($km^2$) | Proportional area | Dominant biota | Category |
|---|---|---|---|---|
| 1 | 0.5 | 0.05 | NIL | A (abiotic) |
| 2 | 2.5 | 0.25 | oligochaetes, some spionids | B (opportunistic) |
| 3 | 5.0 | 0.50 | Macoma Scrobicularia Fucus | C (stable) |
| 4 | 2.0 | 0.20 | Tellina Nepthys | C (Stable) |

## POLLUTION LOAD INDEX

An estimation of the pollution in the estuary may also be obtained from the measurement of the levels of pollutants in some component or components of the estuarine system and this forms the basis for the PLI.

Four components may be monitored:

a. Water: Meaningful monitoring of the water component has been considered expensive and complicated (Forstner, 1980; Tomlinson et al, 1980), and the problems may be exacerbated by presence and interactions of the different water bodies, in particular at the freshwater/brackish water interface (Morris et al, 1982). However, water quality is a commonly used pollution criterion, especially in fresh water (Cook, 1976) and may be justified for example, where migratory fish passage is the sole objective. In selecting intertidal sediment analysis as the tool for assessing the "quality" of the estuarine environment, or conversely its pollution load, there are a number of assumptions. The most fundamental of these is that a dynamic relationship exists between the chemistries of the two major environment components, water column and sediment and all the biota in the estuarine ecosystem.

b. Fauna: the variations in pollutant levels in fauna with species, size, sex, condition, season and tidal height have been extensively discussed by Phillips (1980), and while it is possible to obviate these difficulties, great restriction is placed upon the sampling.

c. Flora: the same drawbacks occur in analysis of the flora as the fauna, although there is some suggestion that the variation is not marked as in animals (Tomlinson et al, 1980).

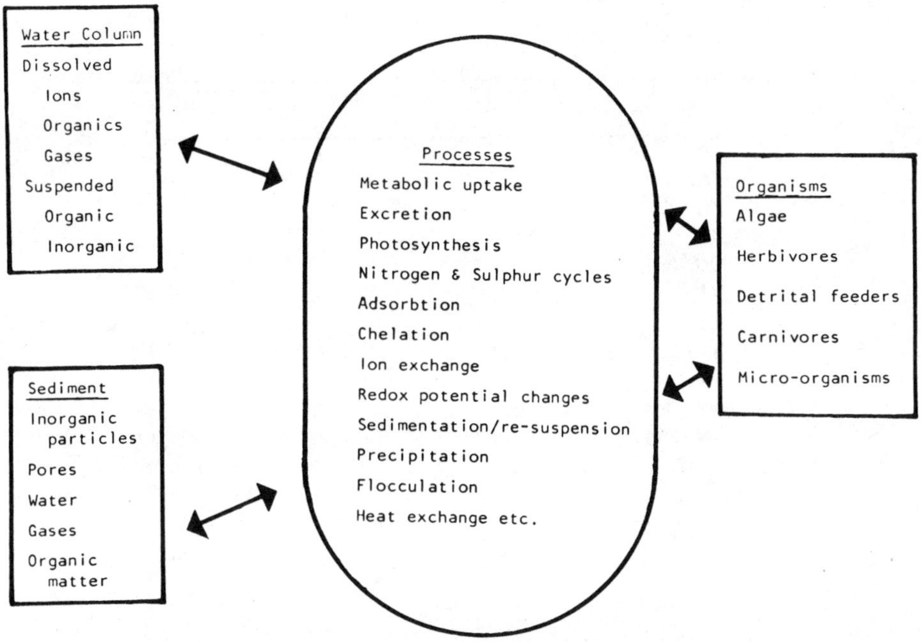


Fig. 1 Diagrammatical representation of estuarine components and processes.

d. Sediment: levels of pollutants, especially heavy metals, have been linked to the organic content of sediments, which in turn is closely related to particle size, as well as to particle size itself (Forstner, 1980; Vivian, 1980).

Of the four, it was felt that the analysis of the sediment would yield the most consistent results. The problem of variation with particle size could be tackled by the analysis of a certain size fraction only.

Other analytical procedures should also be carefully chosen to minimise variation while combing ease and speed of analysis with accuracy. For instance the recommended procedure for heavy metal analysis is digestion in nitric acid, which extracts the major fractions of most heavy metals including any that could be biologically available (Bryan et al, 1980).

In selecting intertidal sediment analysis as the tool for assessing the "quality" of the estuarine environment, or conversely its pollution load, there are a number of assumptions. The most fundamental of these is that a dynamic relationship exists between the chemistries of the two major environment components, water column and sediment and all the biota in the estuarine ecosystem (Fig. 1). Because the components of the system are interlinked, any enrichment of

Table 2 Baseline (B) and threshold (T) values for pollutants (p.p.m.): see also Fig. 2 and text.

| Pollutant | B | T | (T-B) |
|---|---|---|---|
| Cu | 5.0 | 50.0 | 45.0 |
| Fe | 2,000 | 20,000 | 18,000 |
| Mn | 150 | 500 | 350 |
| Pb | 10 | 100 | 90 |
| Co | 5 | 10 | 5 |
| Ni | 5 | 20 | 15 |
| Al | 100 | 500 | 400 |
| Hg | 0.05 | 1.50 | 1.45 |
| As | 5 | 100 | 95 |
| Ag | 0.1 | 2.0 | 1.9 |
| CN | 0.01 | 1.0 | 0.99 |
| $NH_4$-N | 1.0 | ?5.0 | 4.0 |
| PCB | 0.1 | 460 | 450.9 |
| Hydrocarbons | 50 | 600 | 550 |

furthermore, for all manner of practical reasons, cost being the most important (see Goldberg et al, 1978), the minimum number of analyses per sample should be carried out. Thus the battery of tests applied within a given estuary would depend on a previously determined overview of the industrial and urban situation and of other special activities within the catchment or hinterland. The full manual suggests ways of collating this information and deducing a suitable set of analyses to be carried out as routine. The recommended set of analyses are shown in Fig. 2. Added to these might be by way of example: Hg if a chloralkali plant was in the vicinity, hydrocarbons if oil refining or bulk oil handling occurred, specific pesticides which might result from wool processing or rice cultivation and other specific organics according to manufacture or use. Each group of samples therefore yields a series of analytical results, as few as ten or as many as twenty or thirty.

The original form of the PLI was based on the concept of the baseline, and preliminary values for Irish estuaries were published using baseline values for the alga Fucus (Tomlinson et al, 1980). For the reasons outlined above, the component for analysis has been changed to the sediment, and the baseline concept outlined in Jeffrey, Harris et al (1978) enlarged to take into account "threshold" pollutant values. The "threshold" is defined as the lowest concentration associated with unacceptable change. Threshold values were not easy to determine: the threshold values shown in Fig. 2 and Table 2 have been estimated from published data and from surveys of Irish estuaries in which contamination of the sediment has been found in conjunction with depleted biological communities. They must be

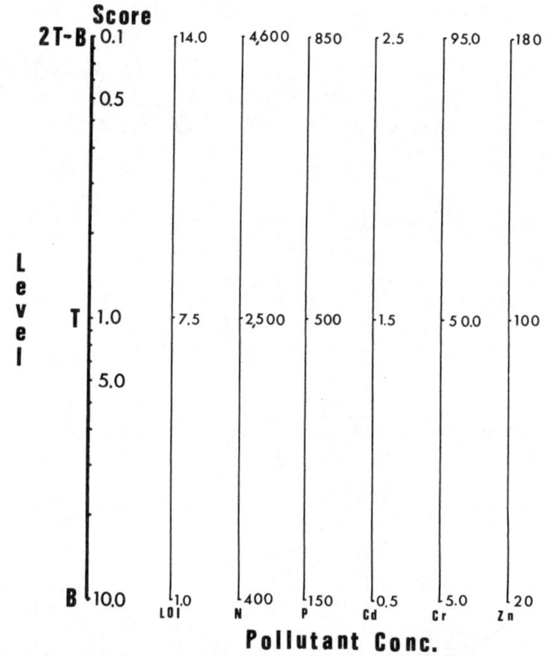


Fig. 2 Scaling of pollutants from baseline (B) through threshold (T) values, and scores for the six recommended mandatory pollutants: organic matter (LOI (%)) total N (N), total P (P), cadmium (Cd), chromium (Cr) and zinc (Zn) (μg/g dry weight sediment). See also Table 3 and text.

an estuary ultimately leads to enrichment of, and hence the possibility of detection in, the estuarine sediment itself (Bryan et al, 1980; Pearson and Rosenberg, 1978). While there are obviously different timescales and mechanisms with the different pollutants, a major practical benefit is the approximately $10^3$ fold amplification in concentration between water and sediment (Forstner, 1980; Vivian, 1980).

A second assumption is that the status of the intertidal zone reflects the status of the estuary as a whole. There are two advantages arising out of the choice of the intertidal zone. From a sedimentation point of view it tends to act as a sink for particulates and their absorbed materials (including pollutants). Moreover, within the intertidal zone environmental differences occur in longitudinal rather than transverse orientations and sampling position within the intertidal cross section is much less important (Jeffrey et al, 1978).

In carrying out an environmental screening of sediments the simplest procedures that yield meaningful results should be adopted and,

Table 3 Example of calculation of the PLI of the Tolka estuary:
a) calculation of site PLI from pollutant loads (ppm unless stated);
b) calculation of estuary PLI from site PLIs.

| | a | | b | |
|---|---|---|---|---|
| Pollutant | conc. | PLI | Site No. | Site PLI |
| Organic C | 18.91% | 0.018 | 1 | $4.31 \times 10^{-3}$ |
| N | 13165.6 | $8.34 \times 10^{-6}$ | 2 | $5.87 \times 10^{-3}$ |
| P | 2785.0 | $2.96 \times 10^{-7}$ | 3 | $3.33 \times 10^{-2}$ |
| Cd | 2.16 | 0.22 | 4 | 1.78 |
| Cr | 34.0 | 2.27 | 5 | 0.26 |
| Zn | 452.0 | $3.98 \times 10^{-5}$ | 6 | 0.568 |
| Cu | 149.7 | $6.09 \times 10^{-3}$ | 7 | 0.788 |
| Fe | 1.52% | 1.84 | 8 | 0.043 |
| Mn | 816.6 | 0.12 | 9 | 0.012 |
| Pb | 495.6 | $4.02 \times 10^{-5}$ | 10 | 1.690 |
| Co | 8.48 | 2.01 | 11 | 0.72 |
| | Site PLI = $4.31 \times 10^{-3}$ | | Estuary Index = 0.1243 | |

regarded as an approximation which should be revised in the light of better information becoming available. The basic unit of the PLI is the difference between B (baseline, unpolluted) and T (threshold, damage) levels based on a logarithmic scale such that doubling the pollution reduces the index value 10-fold. Fig. 2 shows the scaling for the six recommended mandatory pollutants.

The PLI for each pollutant is determined from the formula

$$\text{PLI} = \text{antilog}_{10}\left(1 - \frac{CP - B}{T - B}\right)$$

where CP = pollutant concentration. The scores for each pollutant are summed to give a total site PLI thus

$$\text{site PLI} = (\text{PLI}_1 \times \text{PLI}_2 \times \text{PLI}_n)^{1/n}$$

for n pollutants. The site scores are then summed likewise to give the

$$\text{estuary PLI} = (\text{PLI}_1 \times \text{PLI}_2 \times \text{PLI}_j)^{1/j}$$

for j sites.

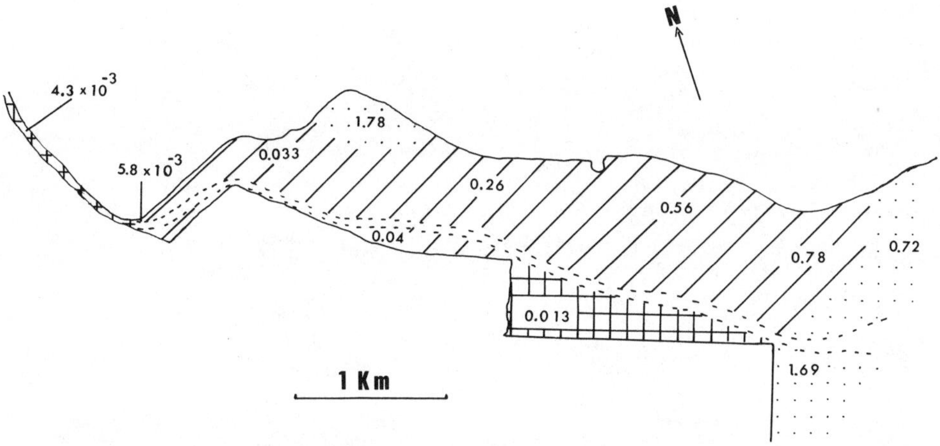


Fig. 3 Tolka estuary, showing biological zones and site PLI values:
#abiotic, //opportunistic, ⁘stable.

A worked example of the PLI for the Tolka estuary is given in Table 3(a) and (b).

From Table 3(a) the PLI of site 1 is $4.31 \times 10^{-3}$ and from Table 3(b) the PLI of the whole estuary is 0.124. The biological classification of the estuary is shown along with the site PLI values in Fig. 3. The outermost areas have been classified as stable with a largely normal fauna, but the rest of the estuary showed signs of faunal depletion (Shannon-Weiner Index $< 1.0$, Wilson, 1983), and the estuary BQI was rather low at 1.83.

The results agree reasonably well with previous work done on the Tolka estuary, which suggested that the levels of some pollutants might be rather high and in places the biota were under stress (Crisp, 1976; Jeffrey et al, 1978; Wilson, 1980, 1983).

The results for some of the other Irish estuaries are shown in Fig. 4. The PLI of some of the east coast estuaries does not differ markedly from that shown by Tomlinson et al (1980) with the exception of the Avoca estuary. The rather higher value in Tomlinson et al (1980) was due to the use of *Fucus* as a variable, and the value shown was for specimens collected just outside the estuary - of necessity, since no organisms were found within the estuary itself. Wilson and McMahon (1981) have demonstrated significant physiological changes in *Mytilus* and *Littorina* with proximity to the Avoca and again noted the absence of specimens within the estuary.

The difference in the Avoca PLI values underlines that one of the virtues of sediment as a subject for analysis is its ubiquity. In polluted estuaries samples may be taken from areas in which no or-

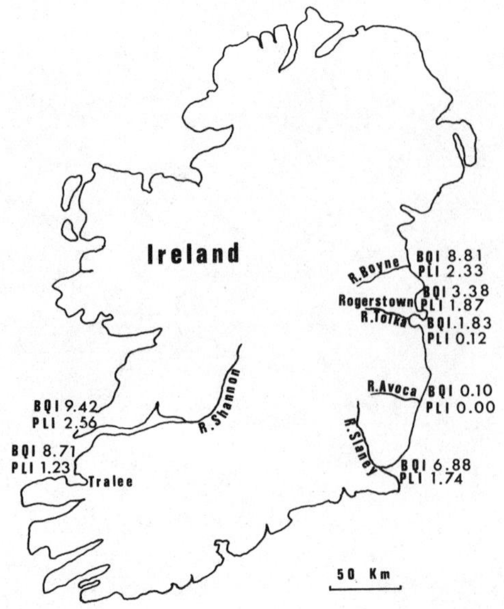


Fig. 4 Estuaries sampled, showing BQI and PLI values for each.

ganisms could survive and this enables a more accurate diagnosis of the nature and magnitude of pollution than analysis of organisms which by their very nature cannot tolerate the worst conditions and therefore may yield a falsely elevated pollution load value.

In the other estuaries on the Irish east coast, Tomlinson et al (1980) publishes PLI values very close to those given here, so it can be concluded that in mildly polluted situations, the choice of analytical material may not be so crucial.

There was a significant correlation between the BQI and the PLI in the estuaries sampled ($r = 0.82$, $P < 0.05$), and Table 4 shows the PLI values associated with the biological classification at site level. Although there was considerable overlap, particularly in the range of PLI values around 1.0, there were significant differences in PLI between the different biological zones. The reasons for this overlap are uncertain, but may be, for example indicative of synergistic effects (both positive and negative) between pollutants.

In conclusion, it appears that the indices presented here have provided a simple comparative means to quantify estuarine quality, and its use could be further extended to indicate trends on a geographical or a temporal scale. While synergism may pose some problems, there would be no obstacle to inserting into the PLI new baseline and threshold values for the synergists when data becomes available, and indeed almost any parameter may be so regarded. It may,

Table 4 PLI values associated with abiotic (A), opportunistic (B) and stable (C) sites and comparison by t-test between classifications A and B and C showing t value (t) and probability (P).

| | | Biological classification | | |
|---|---|---|---|---|
| | | A | B | C |
| PLI | $\bar{x}$ | 0.560 | 1.531 | 3.587 |
| | s.d. | 0.974 | 1.102 | 2.372 |
| | range | 0.0 - 3.10 | $5.3 \times 10^{-3}$ - 3.77 | 0.45 - 10.0 |
| | n | 10 | 43 | 55 |
| | t | | 8.622 | 5.696 |
| | P | | <0.001 | <0.001 |

for example, be possible to determine the appropriate values for the As/Fe and Hg/organic C ratios discussed by Langston (1983). Such data may arise from wider application of the indices by widening the range of situations beyond the rather limited variety found in Ireland in which they have been tested.

Nevertheless, the indices offer the non-specialist a useful yardstick by which to assess for example policy decisions on estuarine quality and it is to be hoped that both administrator and scientist would find them useful, at the very least on the level of promoting thought and discussion!

## ACKNOWLEDGEMENTS

This work was carried out under contract No. 179-77-1 ENV EIR of the EEC Environmental Research Programme. The authors wish to thank all IERP personnel who assisted in the sample collection and handling and Miss G. Moloney for typing the manuscript.

## REFERENCES

Babcock, L.R. and Nagda, N.L. 1975. Indices of air quality. Indicators of environmental quality (Thomas, W.A. (Ed)). Plenum Publishing Corporation, New York 183-197.

Bayne, B.L., Moore, M.N., Widdows, J., Livingstone, D.R. & Salkeld, P. 1979. Measurement of the responses of individuals to environmental stress and pollution: studies with bivalve molluscs. Philosophical Transactions of the Royal Society of London B 286, 563-581.

Bryan, B.L., Langston, W.J. and Hammerstone, L.G. 1980. The use of biological indicators of heavy metal contamination in estuaries. Marine Biological Association of the United Kingdom Occasional Publication Number 1,1-73.

Cook, S.E.K. 1976. Quest for an index of community structure sensitive to water pollution. Environmental Pollution 11, 269-88.

Crisp, D.J. 1976. Survey of environmental conditions in the Liffey estuary and Dublin Bay. Unit of Marine Invertebrate Biology, Marine Science Laboratories, Menai Bridge, Gwynedd, U.K. 74pp.

Eagle, R.A. and Rees, E.I.S. 1973. Indicator species - a case for caution. Marine Pollution Bulletin 4, 25.

Forstner, U. 1980. Inorganic pollution, particularly heavy metals in estuaries. In Chemistry and Biogeochemistry of Estuaries (Olansson, E. and Cato, I., (Eds)). John Wiley & Sons Ltd., New York. 307-348.

Goldberg, E.W., Bowen, U.T., Farrington, J.W., Harvey, G., Martin, J.H., Parker, P.L., Risebrough, R.W., Robertson, W., Schnieder, E. and Gamble, E. 1978. The mussel watch. Environmental Conservation 5, 1-25.

Gray, J.S. 1976. The fauna of the polluted river Tees estuary. Estuarine and Coastal Marine Science 4, 653-676.

Gray, J.S. 1979. Pollution-induced changes in populations. Philosophical Transactions of the Royal Society of London B 286, 545-561.

Hiscock, K. 1979. Systematic surveys and monitoring in nearshore sub-littoral areas using diving. In Monitoring the marine environment (Nichols, D. (Ed)). Institute of Biology, London. 55-74.

Jeffrey, D.W., Harris, C.R., Tomlinson, D.L., Wilson, J.G. and Lyes, M. 1978. The use of organisms and sediments in monitoring estuarine quality. In Coastal Pollution Assessment (Downey, W.K. & Ni Uid, G. (Eds)) National Board of Science and Technology, Dublin. 137-146.

Jeffrey, D.W., Pitkin, P.H. and West, A.B. 1978. Intertidal environment of northern Dublin Bay. Estuarine and Coastal Marine Science 7, 163-171.

Langston, W.J. 1983. Assessment of the distribution and availability of Arsenic and Mercury in estuaries. In Estuarine Management and Quality Assessment (Wilson, J.G. and Halcrow, W.A. (Eds)). Plenum Press, London.

Leppakowski, E. 1975. Assessment of degree of pollution on the basis of macrozoobenthos in marine and brackish water environments. Acta Academia Aboensis (B) 35, 1-89.

Leppakowski, E. 1977. Monitoring the benthic environment of organically polluted river mouths. In Biological monitoring of inland fisheries (Alabaster J.S. (Ed)). Applied Science Publishers, Barking. 125-132.

Morris, A.W., Bale, A.J. and Howland, R.J.M. 1982. The dynamics of estuarine manganese cycling. Estuarine Coastal and Shelf Science 14, 175-192.

Nurnberg, H.W. 1980. Comparative studies on heavy metal pollution and balance in selected regions of European Seas. In Environment and Quality of Life, Second Environmental Research Programme, Commission of the European Communities. CEC, Brussels - Luxembourg. 554-563.

Pearson, T. and Rosenberg, R. 1978. Macrobenthic succession in relation to organic enrichment and pollution of the marine environment. Oceanography and Marine Biology Annual Review 16, 229-311.

Phillips, D.J.H. 1980. Quantitative aquatic biological indicators. Applied Science Publishers Ltd., London. 488 pp.

Sanders, H.L. 1968. Marine benthic diversity: a comparative study. American Naturalist 102, 232-282.

Thomas, W.A. 1975. Indicators of environmental quality. Plenum Press, New York, 275 pp.

Tomlinson, D.L., Wilson, J.G., Harris, C.R. and Jeffrey, D.W. 1980. Problems in the assessment of heavy-metal levels in estuaries and the formation of a pollution index. Helgolander Meeresuntersuchungen 33, 566-575.

Vivian, C.M.G. 1980. Trace metal studies in the river Tawe and Swansea Bay. In Industrialised Embayments ad their Environmental Problems (Collins, M.B., Banner, F.T., Tyler, P.A., Wakefield, S.J. & James, A.E. (Eds)). Pergamon Press, Oxford. 329;342.

Wilson, J.G. 1980. Heavy metals in the estuarine macrofauna of the east coast of Ireland. Journal of Life Sciences, Royal Dublin Society

Wilson, J.G. 1983. The littoral fauna of Dublin Bay. Irish Fisheries Investigations Series B, 26.

Wilson, J.G. and McMahon R.F. 1981. Effects of high environmental copper concentration on the oxygen consumption, condition and shell morphology of natural populations of Mytilus edulis L. and Littorina rudis Maton. Comparative Biochemistry and Physiology 70C, 139-147.

# PROGRESS IN ESTUARINE WATER QUALITY MANAGEMENT: AN OVERVIEW

David W. Mackay

Clyde River Purification Board, River House
East Kilbride, Glasgow, U.K.

## INTRODUCTION

The management of estuaries, and the effective and economic control of pollution in estuaries, has, in common with management generally, five major requirements. These are: clearly defined objectives, technological capability, and a method of measuring performance in cost-effectiveness terms, a firm resolve to meet defined responsibilities in full and possession of effective executive powers. This paper sets out to critically examine how estuarine management has developed in relation to each of these components and to assess future prospects for continuing improvement.

## OBJECTIVES

It may be stated that the overall planning objective is to develop and implement a policy for effluent disposal which secures and maintains estuaries of satisfactory quality at minimum cost to the discharger and to the public purse. Until relatively recently it would have been very difficult to find in the scientific or legal literature a description of what a "clean" or unpolluted estuary should be. The third report of the Royal Commission on Environmental Pollution published in 1972 provided valuable guidance by stating that pollution can be accepted if the estuary is able (a) to support on the mud bottom the fauna essential for sustaining sea fisheries and (b) to allow the passage of migratory fish at all states of the tide.

The seven estuaries studied by the Royal Commission were all industrialised and therefore polluted to some extent. Because of this the objectives can be seen as targets for improvement, rather than acceptable levels to which high quality estuaries can be allowed to deteriorate. More stringent objectives have been set out for undeveloped estuaries by at least one authority (Howells et al, 1978), nevertheless the Royal Commission objectives are generally accepted for application in developed estuaries (Ibid, 1978; Andrews et al, 1983).

Since 1972 considerable effort has gone into providing the specific information necessary to define these broad policy aims more precisely. These definitions may be expressed as Environmental Quality Objectives (EQO), the uses to which the receiving environment is legitimately put, which in turn are achieved by the application of Environmental Quality Standards (EQS), the maximum permissable concentrations of potential pollutants in the receiving environment, and discharge consent conditions on effluents. It is important in practical terms to differentiate between international objectives embodied, for example, in European Community (EC) Directives and the Paris Convention, and local objectives which may be much more specific, and variable both in space and time.

Environmental Quality Objectives are framed to protect the use of saline waters for a number of functions. The most widely applicable objectives can be stated:

(1) As a medium for the growth of food for human consumption. The objective is protection of the human consumer, by restricting levels of a contaminant in any food derived directly or indirectly from saline water, e.g. fish and shellfish, water birds, gulls' eggs, edible seaweed. Objectives are achieved by stipulating either a standard in terms of an allowable level of contaminant in food, or allowable level in water so as to restrict the level in food.

(2) As a habitat for fish and shellfish. The objective is to preserve fish, for angling interests or for protection of an ecosystem, or for general environmental management. Use (1) above may also apply, but only if the fish are eaten by humans. Where possible, a single EQS value will be proposed for all fish and shellfish, and different types of species will only be protected by the setting of individual EQS values where there is strong evidence to suggest a difference in response. This would include the rearing of fish for re-stocking of fisheries. Protection must extend to the most sensitive stages in the life cycle.

(3) As a habitat of other aquatic life and dependent non-aquatic organisms. To protect an organism, its food source or an ecosystem, taking account of sensitive stages in the life cycle. It is usually sufficient to propose a single EQS value for protec-

tion of a wide range of aquatic species and non-aquatic predators, but special, stringent values might have to be set locally for particular spcies or habitats.

(4) For industrial abstraction. Saline waters are usually only abstracted for cooling purposes, with a low quality requirement.

(5) For bathing and contact water sports. For the protection of swimmers and those engaged in water sports.

(6) For aesthetic considerations. The minimum environmental quality necessary to protect public health and prevent visual and smell nuisance.

At the practical management level, the objectives are achieved by the application of environmental quality standards which define the concentration of substance which is acceptable in a body of water in relation to a specified use or uses of that water. In a body of water subject to more than one use the more stringent standard applies.

A good example of the derivation of such standards is the work carried out by the Water Research Centre on behalf of the Department of the Environment to develop EQS for copper, lead, nickel, zinc, chromium and arsenic, six elements on List II of the EC Dangerous Substances Directive (Gardiner and Mace, 1982). However, it is clear that the EQSs proposed at this stage are far from perfect, being derived from what is often sparse data on toxicity and other relevant factors, and they will therefore be subject to review.

If this approach could be applied to the enormous range of potentially toxic materials discharged to estuaries and allowance made for additive and synergistic effects then the search for clearly defined standards might be considered finished. Apart from the sheer magnitude of the task there is one other major difficulty in applying such a system and that concerns the rate of dilution. This will be referred to further in the next section.

In addition to the European Community Dangerous Substances Directive (EEC, 1976a) which has led to the work described above, other EC Directives, notably those on the quality of bathing and shellfish waters, (EEC 1976b, 1979), are worthy of consideration.

Both Directives require designated waters to comply with numerical quality standards within a specified period of time, six years in the case of these Directives. Because of this, designation of presently polluted waters would automatically require capital expenditure on remedial measures within the specified period, so that the standards could be achieved.

The implications for public expenditure were such that Government established stringent criteria for the identification of both bathing waters and shellfish waters suitable for designation. The application of these criteria by the competent authorities, the Water Authorities in England and Wales and the Scottish Development Department in Scotland, resulted in very few designations being made. Only 27 bathing waters were designated in the United Kingdom and none of these were in Scotland. The criteria for designation of Shellfish Waters specified that such designation must have no impact upon capital expenditure, so that the only waters designated were those which already met the required quality standards, or would do so through existing expenditure programmes. Thus Government has largely avoided the consequences of these Directives, where there is always a way out. The way in which the objectives of the Directives have been sidestepped has greatly reduced their value, both as a means of improving water quality and in terms of establishing universally applicable quality standards.

This policy has not, however, been entirely successful. Due to the lack of information on existing water and shellfish quality certain of the designated shell-fish waters do not meet the standards set. In the Clyde River Purification Board area standards for faecal coliforms, zinc and, to a lesser extent, copper have not been met in even the most remote areas.

It may be argued that management objectives must already be sufficiently clear in practice, based on the substantial progress made in cleaning up British estuaries in the last couple of decades. Much of that progress depended on the tide of enthusiasm for environmental protection and improvement which peaked in the early seventies and during which many projects were embarked upon, and commitments to environmental improvement entered into, with little regard to cost.

In my opinion that tide has now receded to the point where those committed to pollution prevention have to rely heavily on the legacy of that upsurge. I have already shown how the EC objectives can, and are, being rendered ineffective in the U.K. context and since legislation should provide a means to achievement I shall deal with it as a tool under the next heading.

## TECHNOLOGICAL CAPABILITY

Here we group everything required for the manager to produce the goods - in this case effective pollution control. Adequate legislative and financial backing, preferably with public and political support, are required. Having these, the necessary manpower and other facilities ca be readily organised.

Legislation may be grouped under two headings - national and international. At the present time control of discharges to estuaries in England and Wales is dependent upon the limited control afforded by the Rivers (Prevention of Pollution) Acts 1951-61 and the Clean Rivers (Estuaries and Tidal Waters) Act 1960. The legislation in Scotland affords similar powers through the Rivers (Prevention of Pollution) Scotland Acts 1951-65. These Acts do not allow control of substantially unaltered existing discharges and in tidal waters are limited in their application to specified estuaries and some areas of inshore waters. Although it is possible to extend these powers by means of "Tidal Waters Orders" this is a difficult and lengthy procedure and although it has been used relatively widely in Scotland there are very few in England and Wales.

The principal difference between Scotland and the remainder of Great Britain lies in the nature of the pollution control authorities. The River Purification Boards in Scotland are independent of the drainage authorities, whereas in England and Wales the powers are vested in the Water Authorities who are responsible for sewage disposal. Although consents for Water Authority discharges in England and Wales are granted by the Secretary of State, and not by the Authorities themselves, there must nevertheless be considerable pressure upon the water quality staff to minimise the degree of treatment imposed upon sewage discharges.

The Control of Pollution Act 1974 when fully implemented will extend pollution control powers to all discharges to tidal waters within three miles of the coast. However the timing and procedures adopted for implementation will greatly reduce the impact of this Act.

Part II of the Act, which deals with water pollution, has been delayed by successive Governments for eight years until it is now imperative that it be implemented to enable the U.K. to meet its international obligations. The Government intends that a five-year programme commencing in July 1983 will introduce the various provisions in a controlled manner.

New discharges to previously uncontrolled waters will then have to receive consent from the pollution control authority, but given the rate of industrial growth in the U.K. at the present time this is hardly likely to make a immediate impact. Additionally, a recent letter from the Government suggests that some future discharges "of a kind or in an area" may qualify for exemption.

Existing discharges will be treated in one of two ways by means of transitional provisions. If they contain List I or II harmful substances they will be granted "deemed consents" which means that a licence will be issued to legalise existing practice and ensure that the discharge can continue unchanged until the discharge consent is determined by the pollution control authority. The vast majority of existing discharges will be subject to "exemption orders" which will obviate the need to issue a licence at all. Only in exceptional cases, to ensure compliance with a particular EC Directive will a licence be required. In earlier consultations, to justify the application of "deemed consents", it was argued tha if there was an especially serious pollution "black spot" then a pollution control authority would be permitted by Government to impose positive controls. It is perhaps noteworthy that in current exchanges no mention is now being made of this important safeguard.

Until these transitional provisions are revoked, the Control of Pollution Act can be no more than a "paper tiger".

On the positive side, the new Act allows wider dissemination of information to the public; in the case of significant discharges, applications ad consents are to be advertised in the press unless the discharger obtains publicity exemption from the Secretary of State. Two years later, registers, containing information on applications and consents, on samples of effluents and water, and on analytical results, together with action taken in consequence of them, will be opened to the public. The following year the purification authority will be required to take action where injurious pollution is caused by a discharge complying with its licence conditions and, eventually, authorities will be empowered to control sewage discharges from boats.

International legislation has already been referred to both as a motive force towards providing clear environmental quality objectives and as a spur towards implementation of our own legislation. The Paris Convention (1974) to which the U.K. is a signatory requires the elimination of pollution by five groups of particularly harmful substances (the black list) and the strict limitation of pollution by certain other substances (the grey list).

Only three EC Directives are of major relevance to coastal pollution. The Dangerous Substances Directive, like the Paris Convention, lists harmful substances into two groups (List I and List II) on the basis of their polluting effects. The Directive does not in itself establish "limit values" or 'quality objectives' for any of the harmful substances but these will appear in subsidiary or 'daughter' Directives. For List I substances, of which there are potentially some 1,500, these will specify uniform emission standards and environmental quality objectives. These are alternative approaches to the control of pollution, the United Kingdom does not favour the arbitrary application of uniform emission standards and has adopted the quality objective approach. This allows effective management of individual discharges to a body of water so that pollution can be eliminated at minimum cost. The EQC approach also ensures that environmentally acceptable levels of pollutants are not exceeded, whereas uniform emmission standards simply require that all discharges receive similar treatment, without regard to ambient water quality.

List I substances which are not yet covered by subsidiary Directives are to be treated in the same way as List II substances meantime. For List II substances, control will be achieved by EQOs and each member state is drawing up its programme for reducing pollution.

Obviously so many substances cannot be tackled equally and simultaneously, so attention is presently focussed on certain active List I substances (cadmium, mercury, and the organohalogens Dieldrin, Aldrin, Endrin, Lindane, DDT and PCBs) and priority List II substances (copper, nickel, chromium, zinc, lead and arsenic).

The Bathing Waters Directive which applied to 27 bathing areas in England and Wales, four of which have derogations, and not at all in Scotland, provides protection for only a small proportion of bathing waters used for recreation and for a minute fraction of the coastline as a whole. The Directive specifies certain physical, chemical and bacteriological standards with which the designated waters must comply by 1985.

The Shellfish Waters Directive, which has already been referred to, requires member states to designate areas of coastal waters needing protection or improvement in order to support shellfish (defined as bivalve and gastropod molluscs, such as scallops and periwinkles, but not crustaceans). These designated waters have to meet certain physical, chemical and bacteriological standards within six years of designation. The Directive is concerned with the quality of water rather than the quality of shellfish (on which a separate Directive is promised) and is not intended to be a public health measure.

The environmental quality standards referred to earlier, arising from the EC Directive on Dangerous Substances, would be very effective if applied strictly, but once the concept of a "mixing zone" is introduced considerable difficulties arise. A "mixing zone" where concentrations can exceed the EQS is essential, some allowance must be made for dilution of the effluent. It is obviously impossible to devise a universally applicable definition of the "mixing zone" as this is highly dependent upon a suite of variables determined by the nature of the discharge and the characteristics of the receiving waters. However, if there is no constraint upon the magnitude of the "mixing zone" then there is a possibility that large bodies of water will be so defined, resulting in unacceptable contaminatin of highly productive inshore waters. There is also the possibility of apparently inequitable application of consent conditions derived from a EQS. For example, one of the principal methods used to improve Britain's estuaries has been the installation of interceptor sewers with trunk mains terminating in a long sea outfall, that is the transfer of pollutants from the head of the estuary to more open waters where more effective dilution can achieved. It is common for effluent to be diluted between 50 and 100 times within a kilometre radius of the point of discharge. Where do we apply the Environmental Quality Standard? Depending on our flexibility in defining the "mixing zone", the acceptable concentration in the effluent could vary by a factor of 1,000.

While there is still much to learn, our knowledge and understanding of what happens to effluents discharged to estuaries has advanced remarkably in recent years. Prediction of distributions and concentrations of substances after periods of time and modelling of the circulation and some of the biochemical and biological reactions of estuarine systems is now accepted as a normal part of selecting an appropriate engineering solution to effluent disposal problems. The monitoring and research programmes carried out over a number of years in several estuaries are increasingly valuable in disentangling the natural variations in the biota from those imposed by man and our understanding of the biology of estuaries is steadily improving. The hardware, that is the instrumentation and sampling devices used in the study of estuaries, has improved in reliability, and often in sensitivity also, over the last decade or two but there have been few introductions of revolutionary technology, the underwater television camera and remote sensing by satellite are worthy of note, and this is somewhat surprising considering the amount of work being carried out. Despite rapid advances in technology much reliance is still placed on judgement and experience in interpreting the data collected.

## COST-EFFECTIVNESS

As estuaries have improved from the extreme condition of being anaerobic, covered with oil and tar, and evil smelling, to perhaps the condition specified by the Royal Commission, there arises the question of how much further we should go, and at what cost. It is fairly straightforward to list the construction and operation costs of sewage treatment works, industrial pollution control plant, long outfalls and so on, but not so easy to evaluate the cost penalty of excluding an industry from a protected area or effectively hindering development of existing industry. The other side of the balance sheet is even more difficult to construct. The principal uses of estuaries which require protection from pollution are as commercial fisheries for human consumption and nursery grounds for such fisheries, for sport such as angling and wildfowling, for recreation, for passage of migratory fish, and for industrial water abstraction. Added to these are the conservation of species generally and rare birds, fish and plants in particular. It could, and almost certainly will, be argued that, given the improvements already effected in British estuaries and the expectation that no major further deterioration will occur, then most of the uses listed above are being achieved at a level where any additional improvement works would be uneconomic. In the fairly long history of water pollution control in England and Wales there seemed to be until recently a philosophy that eventually all receiving waters should be brought up to the highest possible standard, i.e. to a condition where the polluted waters had so far recovered as to be indistinguishable from non-polluted waters. Now, since the guardians of water quality are also the principal dischargers of effluents, the idea that substantial levels of aquatic pollution should be accepted as a normal and permanent feature of parts of our environment is being brought forward and justified in terms of cost-benefit. The problem for those who take a contrary view is that it is virtually impossible to put a cash value on the desire to have the best possible, as opposed to merely a non-harmful environment. Power of public opinion to influence decision-making, perhaps defeating arguments based solely on short-term local, rational and cost-effective considerations, may be weakened by the replacement of locally elected "amateurs" with more "professional" representatives, appointed and paid by central government.

## CONCLUSION

Over the last two decades levels of pollution in several British estuaries have been substantially reduced, the impetus having come from enhanced public concern about the environment, and the initiative having been taken mainly by local government, and water pollution control bodies often lacking in statutory powers. In recent years, international and national legislation has apparently provided, or is about to provide, additional powers and guidance but there is evidence that the will of Central Government to translate the EC Directives and the 1974 Control of Pollution Act into positive action is now lacking. There is sufficient flexibility or loopholes to meet their legal responsibilities by exercises on paper rather than real action. While it may be tiresome to listen to a Scot extolling Scottish systems as opposed to those south of the border, it has to be pointed out that River Purification Boards in Scotland are single purpose authorities with undivided responsibilities, although undoubtedly influenced by national and local economic and other considerations. In England and Wales Water Authorities already appear to place diminished priority on reducing pollution. If it transpires that direction is entrusted to Members who are accustomed to viewing investment in terms of its expected returns in hard cash, then scientists and managers involved with water quality management in estuaries will have to argue the case very convincingly if they are to attract resources into their particular area.

## REFERENCES

Andrews, M.J. Steel, J.E.C. and Cockburn, A.G. 1983. Biological considerations in the setting quality standards in the tidal Thames. _Water Pollution Control_, 82, 52-60.

D.O.E., 1981. _Digest of Environmental Pollution and Water Statistics_. HMSO, London.

EEC, 1976a. A Council Directive on the pollution caused by certain dangerous substances discharged into the aquatic environment of the Community. Council Directive 76/464/EEC, Brussels.

EEC, 1976b. A Council Directive on the quality required of bathing waters. Council Directive 76/160/EEC, Brussels.

EEC, 1979. A Council Directive on the quality required of shellfish waters. Council Directive 79/923/EEC Brussels.

Gardiner, J. and Mance, G. 1982. Determinaion of pollutants in effluents, General Introduction. Water Research Centre Report No. 295-M/1 to the Department of the Environment.

Howells, W.R., Owens, M. and Stoner, J. 1978. Water quality aspects of Welsh estuarine and coastal waters. Journal of the Institute of Water Engineers and Scientists 32, 365-390.

Paris Convention. 1974. A Convention for the Prevention of Marine Pollution from land-based sources. HMSO, London.

Royal Commission, 1972. Pollution in some British Estuaries and Coastal Waters. Third Report of the Royal Commission on Environmental Pollution, HMSO, London.

# THE UK NATIONAL ESTUARINE CLASSIFICATION SYSTEM AND ITS APPLICATION

J.E. Portmann and P.C. Wood

Ministry of Agriculture, Fisheries and Food
Fisheries Laboratory, Remembrance Avenue
Burnham on Crouch, Essex, CMO 8HA, England

## ABSTRACT

An outline is presented of the background to the development of an estuarine classification system used by the UK National Water Council. The basic objectives and several possible approaches which were considered are described. Reasons are given for the selection of parameters and features incorporated in the scheme, which include dissolved oxygen, biological quality and aesthetic quality. The result of the application of the scheme to data collected in 1980 is presented, and the paper concludes with proposals for future modifications and the application of the scheme for management purposes.

## INTRODUCTION

The first survey of river pollution in England and Wales was organised in 1958 by the Ministry of Housing and Local Government. Further surveys were published in 1970 and 1975 (HMSO 1971-1978) by the Deprtment of the Environment (DOE) and involved the use of similar procedures. Following the 1975 survey it was decided that the system of assessment should be made more quantifiable and objective. In view of its role as the coordinating body of the Water Authorities, the task of organising the survey and publishing the results was transferred to the National Water Council (NWC) which also undertook the development of the new river classification system. The new scheme was published in 1978 for use in the 1980 river quality survey (NWC, 1978).

In 1974 the Control of Pollution Act (HMSO 1974a) was enacted giving Water Authorities increased powers in tidal waters, and by

1977, the DOE had concluded that a system of water classification of estuaries and coastal waters should be developed and, if possible, adopted for use in conjunction with the 1980 river pollution survey.

Three main reasons for the conduct of a survey of estuaries and coastal water quality were identified by the DOE, viz;

(a) the provision of a nationally- based classification system would demonstrate to Ministers that their statutory and administrative interests in the state of estuarine and coastal waters were being safeguarded, and that the broad policies to which they had committed themselves were being adequately implemented

(b) a means was required by which reports could be made to the general public, particularly those with strong environmental interests in the quality of estuarine and coastal waters;

(c) a means was required by which reports could be made to the European Commission and Member States of the Community, and to the Paris Commission, on the quality of estuarine and coastal waters, particularly in relation to the effects of discharges of black list and grey list (List I and II) substances to the aquatic environment (HMSO 1974b and EEC 1976).

A number of subsidiary benefits of a classification system were also recognised, for example in assessing the effect on the environment of tidal power schemes, barrages, possible conversion of an estuary to a water storage system, industrial and municipal developments and the assessment of minimum acceptable flows to estuaries. A further advantage was that, by adoption of a uniform method of classification, changes over a period would be apparent and be capable of easy summary, on a common basis.

## DEFINITION OF THE TASK UNDERTAKEN

Responsibility for coordinating the technical aspects of pollution monitoring and assessment procedures in the United Kingdom rests with a number of monitoring management groups (MMGs). These groups were organised under DOE guidance (HMSO 1974c) and report on matters related to their particular field of responsibility. Within this framework estuaries are the joint concern of the Freshwater and the Marine pollutin MMGs. Accordingly, a working party was established, working to these two MMGs, to develop a system of water quality classification for estuaries and coastal waters. One of the authors (PCW) acted as chairman and the other (JEP) as Technical Secretary of the Working Party.

At a very early stage in the discussions a number of problems were recognised in relation to the breadth of the task, and several constraints were identified. In particular it was noted that most coastal waters are of relatively high quality compared to estuarine waters. Thus, any system devised principally for estuarine waters would place virtually all coastal waters in the highest class. It was therefore concluded that a single system would be inappropriate and, as most of the effects of discharges are limited to estuaries, the task was restricted to consideration of estuaries. A major constraint was that the scheme had to be applicable to all estuaries in England, Wales and Scotland, even though this implied many different types of estuary anad the amount of data available for the different estuaries ranged from distinctly estuaries ranged from distinctly limited to extensive.

The main reasons specified by the DOE for needing a scheme of estuarine water quality classification were related to administrative and public relations needs, rather than as a basis for management. Nevertheless, fears were expressed that if an estuary or part of an estuary was classified as other than of the highest quality there would be an implied obligation for it to be upgraded. All those involved in the classification scheme, from working party members to the DOE and the NWC, areed that such arguments would be entirely wrong. Decisions to improve the quality of an estuary can be judged only on the basis of desired estuary use, as specified by the responsible Water Authority, taking due account of local factors such as the physical nature of the estuary, management objectives and availability of funds.

It was therefore concluded that the system of estuary classificaion devised should:

(a) enable a simple assessment to be made of the state of the nation's estuaries, and

(b) enable changes in the quality of estuarine waters to be recorded, particularly in connection with changes in water quality consequent upon measures to reduce pollution.

It was agreed that regard should be paid to estuarine use, e.g. passage of migratory fish, fishing, shell fisheries, wildlife, amenities such as sailing, bathing and appearance etc. and the needs of industry.

## DEFINING AN ESTUARY

Difficulties were encountered in defining estuarine limits for the purpose of the classification. Many attempts have been made to define estuaries according to physical, chemical or hydrographic parameters. None of these were wholly satisfactory for the purpose in hand, or readily applicable to all estuaries for limit definition purposes. It was therefore agreed to regard an estuary, in very general terms, as the transition zone along which the quality of water changes from that of freshwater, characteristic of inland river water, to that of saline water, characteristic of the open sea.

This definition left the question of locating upper and lower boundary limits very vague and allowed the adoption of limits of a purely practical nature. This was considered essential if representative sampling stations were to be established at the upper and lower boundaries. It also allowed due account to be taken of previously defined boundaries such as those laid down by the Clean Rivers (Estuaries and Tidal Waters) Act 1960 (HMSO 1960) and the Rivers (Prevention of Pollution) (Scotland) Act 1965 (HMSO 1965).

In the event, the seaward limits of most estuaries were defined in identical terms to those described in the aforementioned Acts. The main exceptions were the exclusion of areas of open water in large bays such as Morecambe Bay, which are included in the Clean Rivers (Estuaries and Tidal Waters) Act limits, and several extensive areas of coastal waters which are included in the Rivers (Prevention of Pollution) (Scotland) Act. Where Water Authorities were doubtful whether to include such areas of open water, the working party advice was to include them.

Recognising that if the concentration of chloride in water exceeds 200 mg $l^{-1}$ the water cannot be used for irrigation purposes, it was agreed that, as a guide, the landward boundary of an estuary should be that at which the chloride level does not exceed 200 mg $l^{-1}$ at high water of mean spring tides and during low freshwater flow. In practice it was assumed that the boundary would be sited at the nearest fixed sampling site, subject to this being compatible with the river quality survey limits.

## APPROACHES CONSIDERED BUT REJECTED

There were obvious advantages in using a system of water classificaion identical to that used for the river quality survey. However, it was noted that river quality relied heavily on chemical criteria, data for which were unlikely to be available for all estuaries and most of which were of little applicability to the marine situation, e.g. suspnded solids. It was also noted tha the river quality classification system incorporated the EEC Directive related to the Quality Required of Surface Water Intended fo the Abstraction of Drinking Water in Member States (EEC, 1976). This Directive clearly does not apply to estuaries, and existing "estuarine" Directives, e.g. that for Shellfish Growing Waters (EEC 1979), pass or fail criteria. For these reasons it was concluded that there was little point in attempting to use either the system devised for use in assessing river quality or even a modified form thereof.

An examination of published experience elsewhere revealed that the Netherlands authorities had devised a system of classification for their aquatic environment. In this scheme, water quality is designated as one of five classes according to chemical quality, based on dissolved oxygen saturation level, BOD and ammonia content. Points are awarded for different ranges of each determinand and averaged. The final score is compared to a range for each class. It was felt that the summation and range approach was applicable to estuaries, but experience with measurement of BOD and ammonia levels in estuaries has indicated considerable difficulties in obtaining meaningful results. Additionally, BOD values were considered to be of limited and unequal significance along the length of an estuary and especially in different estuaries. For these reasons it was concluded that the Netherlands system was not suitable for direct application to UK estuarine waters.

The classification of estuaries according to their actual or desired use has obvious merits from a control point of view, and it was felt that it ought to be possible to devise a scheme of this nature. However, it is difficult to define accurately, for each of the many chemical parameters, the limiting concentrations for each involved; especially bearing in mind that these would vary according to local circumstances. It was also realised that regions of different use exist within most estuaries and are often either not confined to clearly-defined areas or in areas which overlap and have several uses. It was therefore concluded that a classification system developed purely by use was impracticable. It was, however, considered that it was desirable to incorporate certain aspects of use in the assessment process.

## SCHEME ADOPTED AND WHY

It was decided that there were three ways in which the quality could be judged - on the bases of chemical quality, biological quality and aesthetic quality. Assessment of each of these would be made separately and the separate results summed to give an overall assessment. It was recognised that there was a danger of double counting as, for example, chemical quality can affect both biological quality and aesthetic quality. The system was therefore developed so as to avoid such eventualities as far as is practicable.

## CHEMICAL QUALITY

It is possible to measure, with varying degrees of difficulty, a large number of chemical parameters in estuarine waters. However, the scheme in question had to be applicable to all estuaries in England, Wales and Scotland. It also had to be conducted with the minimum of additional work by the locally responsible water authori-

Table 1 Assessment of chemical quality

| (One value only is chosen) | Percentage | Points awarded |
|---|---|---|
| | 60 | 10 |
| | 40 | 6 |
| Dissolved oxygen exceeds a saturation value of: | 30 | 5 |
| | 20 | 4 |
| | 10 | 3 |
| | <10 | 0 |

ties. It was therefore necessary to avoid inclusion of many possible determinands because relevant data were only available for a few estuaries. This restricted consideration to simple determinands such as BOD and dissolved oxygen, and nutrients including ammonia.

As has already been mentioned, there are practical difficulties in measuring BOD values in waters of differing salinity, and in all but the most polluted saline waters BOD values are low. Use of BOD was discounted for these reasons, but an assessment of dissolved oxygen content was retained as it would to some extent reflect BOD values. The Netherlands system of classification makes allowance for oxygen super-saturation in day-time, but in most British estuaries this is of limited significance and it was decided that daytime measurements should be used without adjustment. It was recommended that the oxygen values used for classification should be obtained under worst conditions averaged with depth and over a tidal cycle.

The classification of water quality according to DO is shown in Table 1. A level of 60% was chosen as the top grade on the basis that estuaries where this represents a worst conditions level would be saturated for substantial periods and it is only below 60% oxygen saturation that estuarine species or migratory fish are likely to experience stress.

The points awarded both here and elsewhere also take account of the practical operation of the scheme i.e. trials were conducted with different points systems on typical estuaries. The points allocations as indicated in this paper were those finally adopted as showing the essential differences in estuarine quality. It should be noted that for this reason the system is intended to be used without interpolation i.e. no intermediate scores are allowed.

Ammonia can be used as a general indicator of pollution but is only measured in a proportion of estuaries in England, Wales and Scotland and it was not expected that reliable measurements could be achieved by the time the survey was due to be completed. Additionally, there did not appear to be a firm understanding of the biological

Table 2 Assessment of biological quality

| Characteristics assessed (Scores under a, b, c and d are summed) | Points awarded |
|---|---|
| a) Allows the passage to and from fresh water of all relevant species of migratory fish, when this is not prevented by physical barriers. (Relevant species include salmonids, eel, flounder, cucumber smelt, etc). | 2 |
| b) Supports a resident fish population which is broadly consistent with the physical and hydrographical conditions. | 2 |
| c) Supports a benthic community which is broadly consistent with the physical and hydrographical conditions. | 2 |
| d) Absence of substantially elevated levels in the biota of persistent toxic or tainting substances from whatever source. | 4 |
| Maximum number of points | 10 |

significance of levels measured, although it is recognised that where ammonia levels are elevated, dissolved oxygen concentrations are usually depressed. Similarly there is no clear quantitative relationship between levels of nitrite, nitrate, phosphate and silicate and detrimental effects applicable to estuaries. It was therefore decided to restrict the use of chemical measurements to dissolved oxygen. The points awarded in Table I recognise the significance of the different levels observed, e.g. the difference between 60% and 40% saturation is believed to be of considerably greater importance than subsequent decreases.

## BIOLOGICAL QUALITY

It was considered that assessment of the biological quality of an estuary involved consideration of at least four separate characteristics, as indicated in Table 2. Points would be awarded if the estuary met the description given; overall biological quality being assessed by summation of points awarded.

The separate assessments indicated in Table 2 were identified for the following reasons.

(a) Passage of migratory fish. Except where other uses are deemed of greater importance, an estuary should allow the passage of all those spcies of migratory fish which ca be supported by the

Table 3 Assessment of aesthetic quality

| Description (One description only is chosen) | Points awarded |
|---|---|
| a) Estuaries or zones of estuaries that either do not receive a significant polluting input or which receive inputs that do not cause significant aesthetic pollution. | 10 |
| b) Estuaries or zones of estuaries which receive inputs which cause a certain amount of aesthetic pollution but do not seriously interfere with estuary usage. | 6 |
| c) Estuaries or zones of estuaries which receive inputs which result in aesthetic pollution sufficiently serious to affect estuary usage. | 3 |
| d) Estuaries or zones of estuaries which receive inputs which cause widespread public nuisance. | 0 |

freshwater reaches. The estuary would fail this criterion if the passage of one or more of the relevant species was seriously impeded by adverse water quality. Thus, certain east coast estuaries, for instance, would not be failed because migratory salmonids do not pass through them, but would be failed if they did not allow the passage of elvers and eels. Similarly, an estuary would not be failed on this parameter if the only impediment to migration was a physical barrier. The main deterrent to migration is usually low dissolved oxygen, and this is reflected in the classification score under chemical quality. However, account should be taken of the possibility that fish might be able to migrate through an estuary with the lowest quality classification if at the appropriate time there is sufficient dissolved oxygen present to allow migration to occur. For example, the lowest dissolved oxygen may occur during the third quarter of the year, whereas elvers and eels migrate during the second and fourth quarters respectively.

(b) Resident fish populations. The population of fish found in an estuary varies according to the physical and predominant hydrographic conditions of the estuary and the season. Where water quality criteria for recreational, commercial or biological grounds are not met, fish populations will probably be reduced, either sporadically or permanently, in numbers or in species; if this occurs the estuary will be regarded as being below standard in this respect.

(c) Benthic communities. As with fish populations, the benthic community of an estuary will vary according to the physical and hydrographic conditions. However, because the benthic community is sedentary, it reflects conditions at a particular location, in contrast to the fish population which is mobile. It is often not easy to determine whether the benthos is healthy or otherwise, although the extremes are readily recognisable. Thus the benthic community may need to show a substantial deterioration before its failure to qualify for points can be stated with certainty.

(d) Persistent toxic or tainting substances. The accumulation of toxic or tainting substances by estuarine organisms may affect either their subsequent acceptability for human consumption, or the viability of populations of sensitive species. The presence of higher-than-background concentrations of persistent chemicals in the biota would not constitute grounds for failing an area of estuary on this parameter, unless the substances approach concentrations which could cause harm to the organisms or render edible species unacceptable for human consumption. Clearly, where, from a knowledge of the nature of inputs to an estuary, there is no reason to expect the accumulation of such substances in the biota, parts of estuaries would be given the highest rating.

It was recognised that the extent of information available on benthos and resident fish populations would vary somewhat from estuary to estuary and that it would not be practicable to carry out additional surveys. Accordingly water authorities were advised to make maximum use of locally available information and to make reasoned assumptions according to the evidence available. It is expected that by the time of the next survey more complete information will be available.

## AESTHETIC QUALITY

Since one of the main uses of the classification scheme was intended to be in an administrative and public relations context, aesthetic aspects of estuarine quality are an important factor. Because there is no generally accepted scale for assessing aesthetic quality, it must inevitably be subjective. Despite the difficulties inherent in a sbjective assessment, after a process of trial and elimination using a number of reasonably well understood estuaries, four broad divisions were developed and scores allocated as in Table 3.

In assessing aesthetic quality the following factors are considered¦ smell, colour, debris, oil, recognisable sewage solids, and effects from discharge of domestic or industrial effluent. The assessment takes account of natural turbidity, algal growth and the frequency with which floating debris is encountered.

Table 4 Overall classification

| Classification | No. of Points | Description of Quality |
|---|---|---|
| Class A | 30-24 | Good |
| Class B | 23-16 | Fair |
| Class C | 15-9 | Poor |
| Class D | 8-0 | Bad |

Consideration was given here to the use of *E. coli* as a basis for measuring the aesthetic quality of a estuary. After consideration, this was rejected on the basis that *E. coli* concentration is a indicator of faecal pollution and that the aesthetic impact of sewage on an estuary essentially rested in the faecal solids, which were already included. Additionally, the extra effort of sampling and measuring *E. coli* concentrations was thought not to be warranted in view of the difficulty of interpreting the significance of *E. coli* in water.

## OVERALL ASSESSMENT

Having made assessments of chemical, biological and aesthetic quality and awarded points accordingly, a means was required of dividing the summed totals into a broad classification. Various possibilities were considered for estuaries, and tests with a number of examples were made. It was concluded that five divisions gave a spurious impression of precision and that three divisions would not allow small improvements to be reflected. Four divisions, as used for existing freshwater classification, were therefore adopted as indicated in Table 4.

When considering the relative importance of chemical, biological and aesthetic qualities it was concluded that, as the three assessments are inter-related in a complex manner, there was no justification for biasing the scores in favour of one or other parameter, although for practical reasons the method of sub-division on a quality scale was done differently for each assessment.

Because the scores apply to water quality in overall terms, it was considered that simple descriptive terms were undesirable additions to Class A etc. However, for administrative reasons the terms good, fair, poor and bad were adopted by the NWC.

## RESULTS

The system outlined above was used for the 1980 survey and the results have been published by the NWC (1981) Table 5, taken from

Table 5 Estuary quality by Water Authority, 1980 (from NWC (1981))

| Water Authority | Estuary length by new estuary classification scheme (km) | | | | |
|---|---|---|---|---|---|
| | A | B | C | D | Total |
| North West | 220 | 90 | 71 | 69 | 450 |
| Northumbrian | 41 | 43 | 7 | 43 | 134 |
| Severn Trent | 11 | 72 | 0 | 0 | 83 |
| Yorkshire | 5 | 17 | 18 | 0 | 40 |
| Anglian | 453 | 52 | 7 | 2 | 514 |
| Thames | 49 | 89 | 0 | 0 | 40 |
| Southern | 307 | 60 | 13 | 380 | |
| Wessex | 44 | 50 | 14 | 0 | 108 |
| South West | 330 | 17 | 2 | 0 | 349 |
| Welsh | 321 | 89 | 4 | 0 | 414 |
| <u>Shared es-tuaries</u> | | | | | |
| Humber | 29 | 34 | 0 | 0 | 63 |
| Severn | 18 | 28 | 0 | 0 | 46 |
| Survey Totals | 128 | 641 | 136 | 114 | 2719 |

Notes:
1) Water Authority figures exclude shared estuaries
2) North West WA figures include 6 km of tidal canals
3) Lengths are rounded independently to the nearest km and may not sum to totals
4) For details of estuary classification scheme see Table 4.

data given in that publication, indicates that of the approximately 2,700 km length of estuaries in the England and Wales survey, about 67% was regarded as Class A (good), 24% as Class B (fair), 5% as Class C (poor) and 4% as Class D (bad). These figures cannot be compared with the outcome of previous surveys because the boundaries used were different and previous surveys dealt with tidal rivers rather than with estuarine waters proper. Assessment of deterioration or improvement will therefore have to await the next survey due in 1985.

As the survey was conducted on the basis of individual Water Authority assessments, an important feature of the results for 1980 is the extent to which they reflect in relative terms the quality of estuaries, especially the major estuaries, in England and Wales. Whilst it is possible that some optimism may have influenced the assessments, this seems to have been a fairly uniform feature and, as the examples in Figure 1 indicate, the state of the major estuaries does seem to have been reflected in reasonably accurate comparative terms.

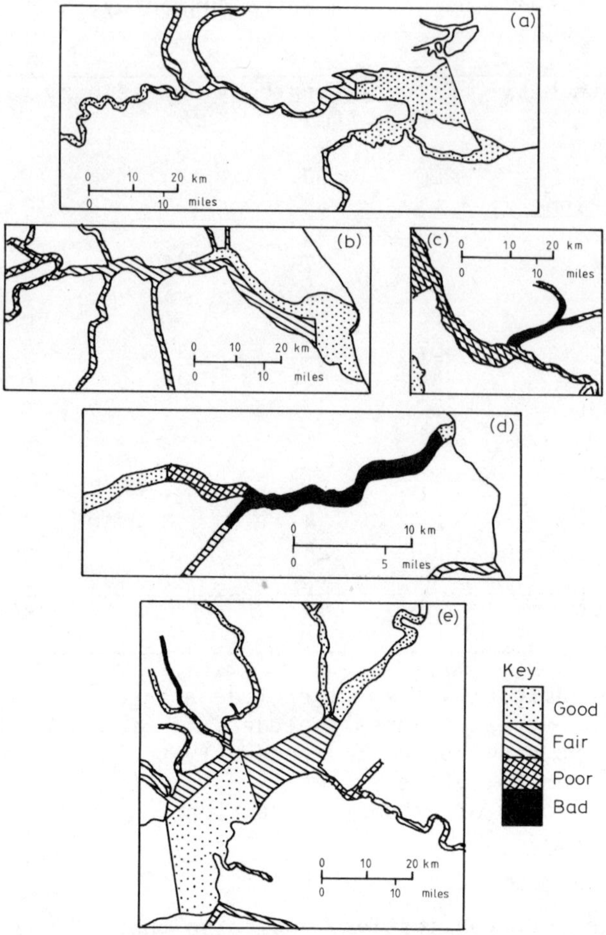


Fig. 1 Results of the 1980 survey of the quality of five estuaries in England and Wales (NWC, 1981): (a) Thames, (b) Humber, (c) Mersey, (d) Tyne, (e) Severn

FUTURE

Possible modifications It is obviously too early to consider changes to the system, as one of the original objectives was to use the scheme for assessment of changes wit time, both in terms of overall estuarine quality and in relation to particular estuaries or areas of estuaries. However, it has already been suggested that a group be reconvened to consider what improvements are desirable and/or practicable. Two possible needs, apparent from the 1980 survey, are¦

(a) the development of a means indicating the extent to which assessment has been subject to allowance for natural features such as high turbidity or slow flushing rates, and

(b) some means of subdividing Class A so that differences can be shown between the very high quality estuaries, e.g. the Wash and Morecambe Bay, and the good but slightly inferior areas of the Humber and Thames.

Use as a management tool. The extent to which refinements of the type outlined above can be incorporated will affect the value of the results in management terms. One of the weaknesses of the existing system of classification is that it does not relate water quality to either the medium or the long-term use of each area of water, although the present scheme describes to some extent existing use. It is considered that with experience, and when greater funds can be made available for estuarine improvement, it might be possible to incorporate an indication of the extent to which the observed water quality matches up to desired future use. This would allow a clear distinction to be made between areas where water quality failed to match an intended use, and other areas where, for good management reasons, it was not necessary to improve water of relatively low quality because that quality was already adequate for intended use. An example of the latter is found in the upper reaches of the Thames Estuary where water quality could only be improved by very substantial investment but without significant extension of water use.

## REFERENCES


EEC, 1975. Council Directive of 16 June 1975, concerning the quality required of surface water intended for the abstraction of drinking water in the Member States, 75/440/EEC. Official Journal of the European Communities L194, 26-31.

EEC, 1979. Council Directive of 30 October 1979 on the Quality required of Shellfish Waters. Official Journal of the European Community L281, 47-52.

EEC, 1976. Council Directive of 4 May 1976 on Pollution Caused by Certain Dangerous Substances Discharged into the Aquatic Environment of the Community. Official Journal of the European Community L129 23-29.

HMSO, 1960. Clean Rivers (Estuaries & Tidal Waters) Act 1960 London.

HMSO, 1965. Rivers (Prevention of Pollution) (Scotland) Act, 1965, Ch.13 London.

HMSO, 1971. Report of a River Pollution Survey of England & Wales 1970 Vol.1. Department of the Environment & Welsh Office, London.

HMSO, 1974(a). Control of Pollution Act 1974 Chapter 40, 136 pp.

HMSO, 1974(b). Convention for the Prevention of Pollution from Land-based Sources, Paris 4 June 1974, Cmnd 5803. London.

HMSO, 1974(c). The Monitoring of the Environment in the United Kingdom. A report by the Central Unit on Environmental Pollution, Department of Environment. Pollution Paper No. 1, 66 pp.

HMSO, 1978. River Pollution Survey of England and Wales, updated 1975. Department of the Environment and Welsh Office, London.

NWC, 1975. River water quality: the next stage. Review of discharge consent conditions. National Water Council, London, 10 pp.

NWC, 1981. River quality: The 1980 survey and future outlook. National Water Council, London 39 pp.

# THE IDENTIFICATION AND EVALUATION OF ENVIRONMENTAL QUALITY IN SOUTHAMPTON WATER, U.K., USING LIMITED MANPOWER RESOURCES

P.G. Soulsby, D. Lowthion and M.C.M. Houston

Southern Water Authority, Resource Laboratories
Otterbourne, Hampshire, U.K.

## ABSTRACT

A low level sampling programme determined by the manpower resources available has been used to determine the present environmental quality of a industrialised outer estuary - Southampton Water, U.K.

Surveys were made over 4 years of the benthic infauna and the sediment metal levels and over 3 years of the water column nutrient and metal levels, the programme occupying 65 man days per year.

The data obtained is examined to determine the level of change that can be detected and the level of information provided. The data from each parameter studied was capable of detecting gross differences both temporally and spatially but such survey levels could not detect more subtle differences or trends. The water quality status of Southampton Water is sufficiently described by such surveys but other studies are required to establish cause and effect relationships.

## INTRODUCTION

The responsibility for the water quality management of estuaries and coastal waters of Hampshire, Southern England is held by the Southern Water Authority. Present legislation, the future implementation of the Control of Pollution Act 1974 and the various National and International obligations require a significant effort to establish the water quality status of these areas and to provide the required information.

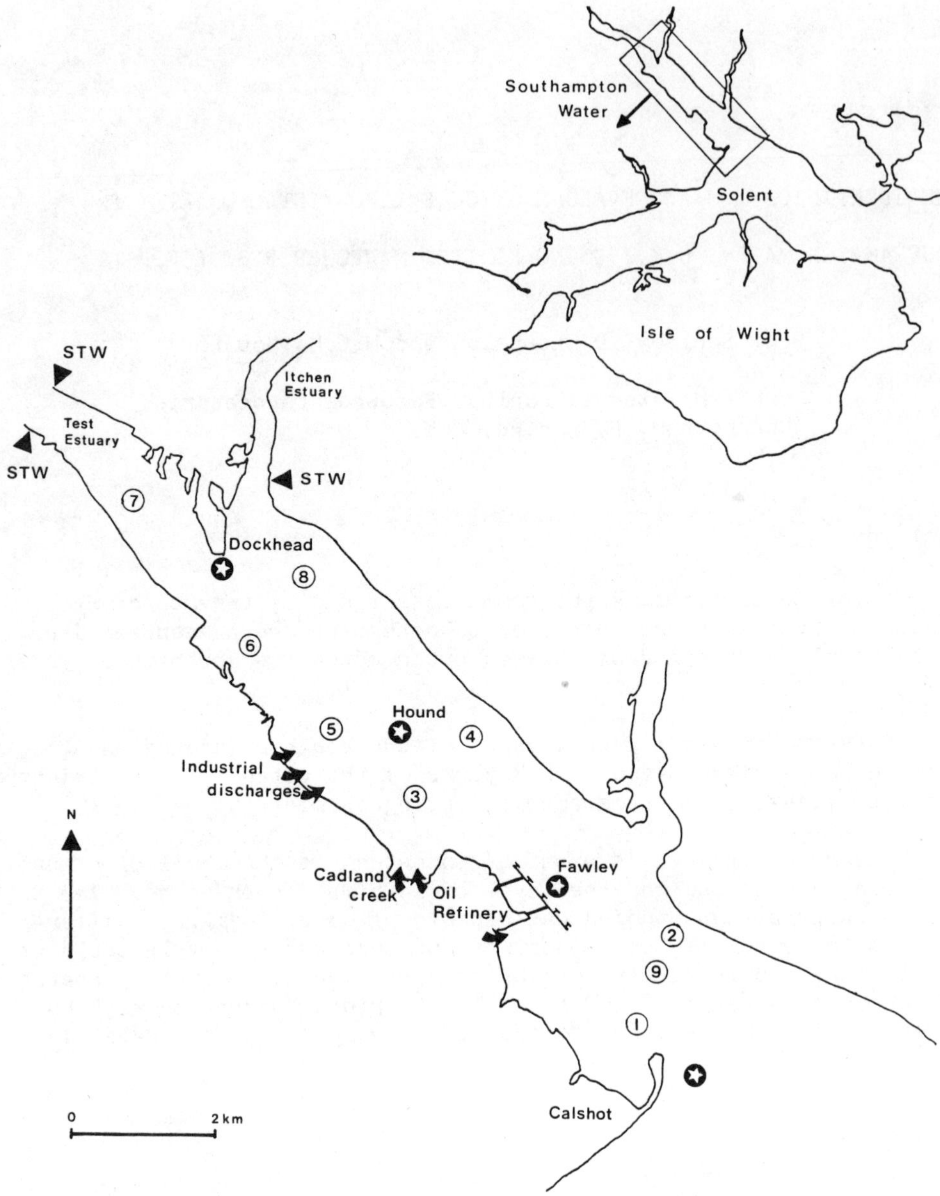


Fig. 1 Southampton Water. Position of industrial and major sewage discharges, water column sampling points, and benthos sampling points.

To obtain the information needed can demand a considerable effort in terms of manpower and equipment resources. These demands have to be satisfied in the context of the Water Authority's entire sphere of activities which itself is established within the framework of prevailing economic conditions. In the current climate therefore, the manpower and resources allocated to marine environmental programmes are likely to be limited and the effort that can be devoted to one particular area even more so. This paper describes the identification and evaluation using a manpower input of approximately 60 man days per year, of the environmental quality of Southampton Water, U.K. (Fig. 1) which receives large industrial and domestic discharges.

## Southampton Water

Southampton Water (Fig. 1) is a shallow, linear body of water approximately 10 km long and 1 km wide. The waters are shallow dredged to 10 m in the shipping channel and 1-8 m elsewhere with some extensive intertidal mudflats and *Spartina* marsh. The tidal regime has been well documented (Webber, 1973, 1980) having a characteristic double high water and a tidal excursion of up to 2.5 km. Two major rivers enter the upper end of the water, the Rivers Test and Itchen which have narrow, stratified estuaries but their effect on the salinity regime of Southampton Water is less marked. Surface salinities rarely fall below 25‰ at its head (Dockhead) and bottom salinities 31‰ at its head and 33‰ at its mouth (Calshot). The sediments tend to be composed of muds with high percentages of the silt (<63 µm) fraction particularly on the western side.

Southampton Water receives the discharges from both domestic and industrial sources including cooling water (Wright, 1980). The bulk of the $5.8 \times 10^5$ $m^3$/day of industrial effluent including the largest oil refinery in the U.K., discharges directly to the estuary, mostly from the western shore.

The estuary supports considerable recreational activities particularly on the eastern side, has a commercially important clam fishery, is an important route for migratory salmon and sea trout and the adjoining Solent supports the largest natural oyster fishery in Europea waters.

The water quality status of Southampton Water was reviewed by Raymont (1972) who concluded that it was not grossly polluted in terms of concentrations of primary nutrients, some metallic ions and dissolved oxygen levels. However, he argued that further increases in quantities of nutrients, and some metallic ions, might have deleterious effects. In general it can be stated that at that time no studies to determine the overall effects of water quality in the estuary had been undertaken nor were there any systems established to monitor changes in the water quality present. Investigations on nutrient levels (Phillips, 1980), some metals in sediments and bivalve molluscs (Romeril, 1974, 1979) and in a variety of organisms (Knap, 1972; Leatherland and Burton 1974), were however in progress and a considerable number of research projects on a variety of parameters were initiated in the mid-1970's (N.E.R.C., 1980).

The Southern Water Authority approach to measuring water quality effects in Southampton Water.

After taking account of previous studies and the methods available, it was considered that the initial approach would consist of 3 investigations. These could be undertaken within the framework of resources available and would allow an improved statement on the water quality status of the estuary to be made and provide a basis from which trends in important parameters could be determined. The three investigations were to be an examination of the benthic infauna, the distribution of sediment metals and an examination of the nutrient and metal concentrations present in the water column. An important component in each of these studies was to determine the variation inherent in the system and its consequent effect on trend determination. Facilities were not available to analyse the important parameter of sediment hydrocarbon values. A detailed survey of these, however, was made by Knap (1979) and the sediment metal analyses would provide additional information to complement these.

The effects of water quality on an estuarine system can be described in numerous ways but without determining biological effect a realistic standard against which to determine necessary changes in inputs or to measure the success of changes is missing. Biological monitoring provides the only valid basis for measuring the sum and product of effects produced by the many substances which might be input into the estuary, (Mackay. 1975). The biotic component chosen for study was the benthic infauna because of their ability to integrate effects over time which other components of the biota cannot (Gray, 1979). Apart from specific studies (George, 1964; Barnes, 1973; Mitchell 1974; Holmes and Coughlan 1975; Hibbert 1976; Coughlan, 1977), mainly on the intertidal and fouling fauna, the benthos of Southampton Water had not received overall coverage although the Solent to which it joins was the subject of a comprehensive dredge survey (Barnes et al, 1973).

Although examination of the benthos is likely to provide detailed information on the delineation of zones of effect, it is an approach which can require a considerable time input. The survey must be approached in the knowledge that it must be long term to describe the natural variations that will occur. The number of sampling points chosen was confined to 8, (Fig. 1), to be sampled annually for a minimum of 5 years.

Sediment samples for metal analysis can be collected simultaneously with the benthic samples. The analysis of sediments for metal levels is relatively easy and can provide valuable information on recent metallic inputs. The programme was therefore to consist of annual samples from the 8 benthic sites; analysis to be based on the $<125\ \mu m$ $>65\ \mu m$ sediment fraction. The metals to be analysed were Cu, Zn, Pb, Cr, Cd, Ni and occasionally, Hg.

The third aproach to identifying the water quality status of the estuary and also as part of an Anglo-French Channel Study, consisted of a series of 0.5 m depth water column samples taken over a 3 year period at high water to determine nutrient and metal levels. Four mid-channel sampling points were chosen in Southampton Water (Fig. 1), with a further 5 in the adjacent Solent. Samples were taken approximately fortnightly during the summer (April - September) and monthly during the winter (October - March).

Water column measurements, apart from dissolved oxygen, are perhaps the least satisfactory method of determining the quality status of an estuary but particularly in the case of nutrients are readily measurable and can provide a monitoring system which should respond to input changes. The objective of the study was to establish the number of sampling points required for a long term monitoring system by delineation of areas of similar status and to determine the number of samples required in a permanent monitoring system to detect a realistic change.

## RESULTS

### Benthic Infauna

The detailed results obtained from the benthos survey have been described elsewhere (Houston et al, in press). Briefly, the sites could be reasonably divided into 3 groups each characterised by a particular faunal assemblage (Fig. 2).

Group 1 consisted only of sites 3 and 5, closest to the refinery and delineating an area of approximately 85 hectares. The second group of sites 1, 6 and 7 occurred on either side of the group 1 sites and the group 3 sites, 2,4 and 8 were present on the eastern side of the estuary. The site groups and their associated species were validated using both ordination ad classification techniques. The 5 years of sampling demonstrated some temporal and spatial variability confirming the value of long term approach to evaluate benthos patterns.

The analysis of the biota results 1975-1980 suggest that the area on the western side, sites 3 and 5, has a severely reduced fauna

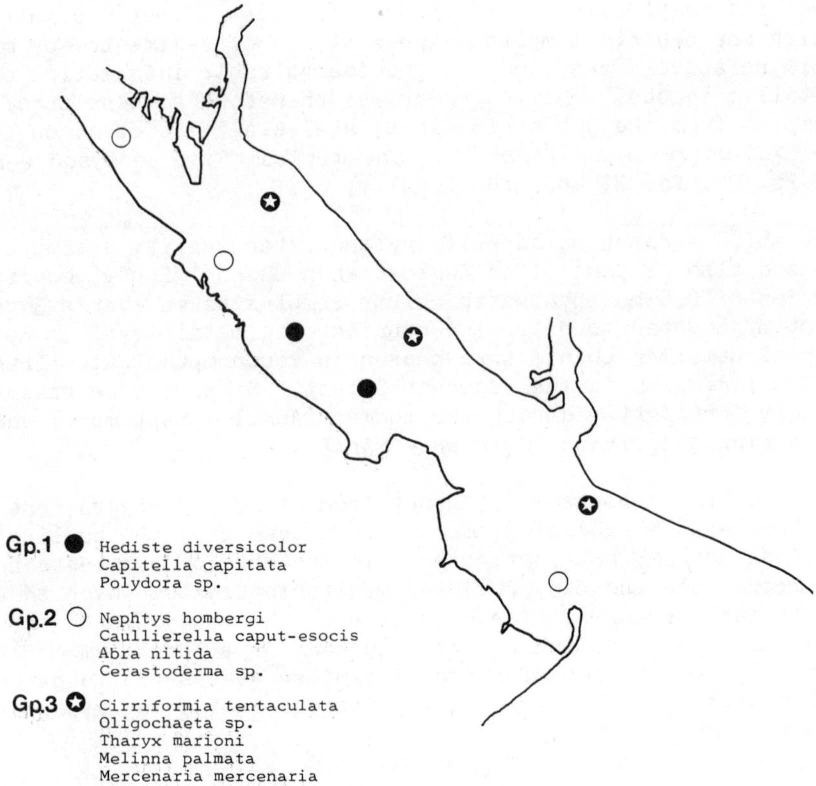


Fig. 2 Distribution of the invertebrate assemblages dominating the infauna at the eight sampling sites 1975-80.

related in some way to the industrial discharges present. Mitchell (1974) speculates that the absence of clams Mercenaria mercenaria may be partly attributable to these effects and Houston et al (in press) conclude that the toxicity threshold of unidentified anthropogenic substances in the area is reached for the larvae of all but the most opportunistic species in the estuary. They are, however, unable to reach a conclusion on the existence of any water quality modification to the fauna of the other 2 groups. More exhaustive surveys by Levell (1979) and Oyenekan (1981) confirm the existence of the faunistically poor area and do not identify other areas affected by poor water quality. McLusky (1982) was able to differentiate different zones of intertidal fauna in the Forth estuary affected by refinery effluent but only broad differences are apparent in the sub-tidal reaches of Southampton Water.

It can be concluded from an examination of the benthos that an area of approximately 85 hectares on the western side of Southampton

Table 1 Sediment trace metals in Southampton Water. Mean concentrations (mg/kg dry weight) ±2 standard errors, 1975-80

| Site | Cu | Ni | Cr | Pb | Zn | Cd | Hg |
|---|---|---|---|---|---|---|---|
| SW1 | 61±29 | 16±7 | 15±7 | 25±12 | 58±18 | 1.3±0.3 | 0.9±1.6 |
| SW2 | 39±12 | 13±7 | 9±2 | 32±14 | 53±15 | 1.2±0.4 | 0.8±0.8 |
| SW3 | 241±74 | 33±7 | 36±16 | 47±11 | 135±31 | 2.0±0.8 | 0.9±0.7 |
| SW4 | 79±52 | 25±5 | 26±7 | 73±18 | 140±50 | 3.1±1.1 | 0.9±0.3 |
| SW5 | 422±136 | 30±6 | 44±20 | 82±34 | 165±45 | 1.9±0.8 | 0.5±0.1 |
| SW6 | 150±38 | 24±6 | 21±6 | 63±11 | 125±20 | 1.9±1.1 | 0.7 |
| SW7 | 65±20 | 22±12 | 16±10 | 49±11 | 75±27 | 1.2±0.3 | 0.8±0.2 |
| SW8 | 58±25 | 20±13 | 13±3 | 57±18 | 95±30 | 2.4±1.3 | 1.3±2.0 |

Water has a very reduced fauna, probably attributable to the industrial discharges; but that any effects on the fauna of the remaining area cannot be distinguished. These therefore are non existent or are not at a level which is readily distinguishable by these methods.

## Sediment Metals

The sediment metal results are shown in Table 1. Taking the results as a single mean for each site enables patterns of distribution to be identified. The data for individual years was subject to some variation at each site with coefficients of variation for each metal and each site varying between 30 - 75%, no site or metal showing any consistency. Site 4 showed a trend of increasing copper and zinc levels subsequently shown to be attributable to a nearby discharge but otherwise no trends in the data over the sampling period were apparent.

The distribution of sediment metals (Table 1) is characterised by peak levels being found in the mid-reaches of the estuary. This applies for both the western side sites 7,6,5,3,1 and the eastern side sites 8,4 and 2, with the highest values occurring at sites 3 and 5 on the western side and 4 on the eastern. However, there are some minor distribution differences which are useful indicators of the metal sources. Multiple correlation of the data demonstrates that the strongest links are between copper, chromium and zinc with lead linked to the latter only. Copper, chromium and zinc have the highest enrichment rates at site 5, with values of 6.8, 3.0 and 2.0 compared to the means of sites 7 and 8 which represent the values for the head of the estuary. Lead has a enrichment ratio of only 1.5 and has the highest loss, 50%, between the head and mouth of the estuary (mean sites 7 and 8 compared to mean sites 1 and 2), similar to zinc, 40%, and higher than chromium and copper, 20% each. Enrichment of the

Table 2 Seasonal means ±2 standard errors of nutrients mg/l and annual means of trace metals μg/l in surface waters at high water. Summers 1977, '78, '79. Winters 77-78, 78-79. Sites underlined not significantly different from each other; p=0.05; S.N.K. test.

| Sampling Period | Dockhead | Hound | Fawley | Calshot |
|---|---|---|---|---|
| Summer $NO_3$ | 0.392±0.085 | 0.266±0.065 | 0.231±0.057 | 0.176±0.051 |
| Winter $NO_3$ | 0.483±0.108 | 0.399±0.112 | 0.271±0.088 | 0.220±0.078 |
| Summer $NH_3$ | 0.115±0.027 | 0.103±0.028 | 0.110±0.034 | 0.063±0.014 |
| Winter $NH_3$ | 0.212±0.055 | 0.188±0.054 | 0.140±0.062 | 0.081±0.026 |
| Summer $PO_4$ | 0.058±0.008 | 0.050±0.008 | 0.051±0.011 | 0.046±0.008 |
| Winter $PO_4$ | 0.074±0.022 | 0.071±0.018 | 0.058±0.021 | 0.06±0.016 |
| Annual Cu | 3.55±0.52 | 2.48±0.40 | | 2.41±0.51 |
| Annual Ni | 1.63±0.33 | 1.16±0.28 | | 1.60±0.42 |
| Annual Pb | 2.50±0.54 | 1.80±0.57 | | 1.71±0.55 |
| Annual Zn | 7.85±1.21 | 6.12±1.20 | | 5.73±1.36 |
| Annual Mn | 9.57±0.66 | 7.93±1.12 | | 5.98±1.36 |
| Annual Cd | 0.37±0.10 | 0.22±0.06 | | 0.28±0.62 |

sediment for copper and chromium occurs in mid-estuary from the industrial discharges on the western side. These are also probably a source of zinc but both zinc and lead are enriched by inputs at the head of the estuary, their sediment levels falling away further towards the mouth than copper or chromium. Apart from maximum levels of copper at site 5, the level of sediment metals in the estuary is not high compared with others in the literature for British estuaries.

The sampling approach used required that a change at any site would have to be large before it could be considered as significant at 95% at a low contamination site, site 2, would require an increase of 115% and similarly a significant decrease at a high contamination site, site 5, 104%. For zinc at the same sites the changes would have to be 101% and 87% respectively. However, the levels of metals at enriched and non-enriched sites are well known and although subtle or large short term trends will be missed, substantial and therefore important changes over a period of years will not.

## Water Column

The results of the water column sampling are shown in Table 2. The data for nutrients is separated into a winter period October - March and a summer period April - September. The summer data is a mean of 3 summers' results 1977, 1978 and 1979 and the winter data 2 winters, 1977 - 78 and 1978 - 79. There are significant differences

Table 3 The number of samples required to detect a 25% or 50% change in the mean values of water column nutrient and metal determinands (p=0.05) at the head and mouth of Southampton Water.
ND = Not detectable

| Station | Sampling Period | Determinand | % Change in mean to be detected 25 | 50 |
|---|---|---|---|---|
| Dockhead | Summer | $NO_3$ | 127 | 8 |
| Dockhead | Winter | $NO_3$ | 53 | 3 |
| Calshot | Summer | $NO_3$ | ND | 18 |
| Calshot | Winter | $NO_3$ | ND | 11 |
| Dockhead | Summer | $NH_3$ | 263 | 9 |
| Dockhead | Winter | $NH_3$ | ND | 4 |
| Calshot | Summer | $NH_3$ | 171 | 9 |
| Calshot | Winter | $NH_3$ | ND | 9 |
| Dockhead | Summer | $PO_4$ | 173 | 3 |
| Dockhead | Winter | $PO_4$ | ND | 4 |
| Calshot | Summer | $PO_4$ | ND | 7 |
| Calshot | Winter | $PO_4$ | ND | 5 |
| Dockhead | Annual | Cu | 22 | 4 |
| Calshot | Annual | Cu | 117 | 9 |
| Dockhead | Annual | Pb | 143 | 9 |
| Calshot | Annual | Pb | ND | 29 |
| Dockhead | Annual | Zn | 25 | 4 |
| Calshot | Annual | Zn | 548 | 12 |
| Dockhead | Annual | Ni | 96 | 9 |
| Calshot | Annual | Ni | ND | 17 |
| Dockhead | Annual | Cd | ND | 16 |
| Calshot | Annual | Cd | 222 | 11 |

between seasons but no differences between years, except for phosphate at the seaward end in summer 1979 which was significantly lower than in the two previous years. The water column total metal results did not vary seasonally and there were no significant differences between years.

The nutrient data demonstrates an apparent concentration gradient down Southampton Water for ammoniacal and nitrate N. The nitrate mainly originates from the 2 chalk derived rivers, the Test and Itchen but the dilution curve for ammonia is interrupted demonstrating the input of ammonia from the industrial discharges midway along the estuary. Levels of phosphate are apparently well buffered and demonstrate no patterns of distribution.

The water column metal levels show no significant trends down the estuary except for copper which is higher at Dockhead than at

Table 4 Annual man days and total costs (1981) of S.W.A. investigations into Southampton Water

| Investigation | Annual Man Days | Total Costs (Pounds) |
|---|---|---|
| Benthos studies | 30 | 1430 |
| Sediment studies | 2 | 100 |
| Water column studies | 33 | 3500 |
| Totals | 65 | 5030 |

Calshot. The levels of lead, cadmium and nickel in the estuary and in the outer Solent are similar but those for zinc are significantly higher at Dockhead than in the Solent.

Levels of all three nutrients analysed were similar to or exceeded those quoted for similar salinities on the N.E. coast of the United States (Maurer et al, 1978). The seasonal differences present for both ammonia and nitrate represent biological uptake and considerable phytoplankton activity is supported during the summer months calculated as contributing 23% of the annual organic input to the systems (Knap et al, 1979). The peak bloom generally appears in late June/July though this can vary (Souza Lima and Williams, 1978), with chlorophyll maxima at the head of the estuary exceeding 50 μg/l mostly contributed by the photosynthetic ciliate <u>Mesodinium rubrum</u> (SWA unpublished). Considerable quantities of the macroalgae <u>Ulva</u> and <u>Enteromorpha</u> are present in some of the intertidal mudflats. Both red coloured water (<u>Mesodinium</u>?) and large quantities of <u>Ulva</u> spp were reported as present at the beginning of the century (Royal Commission 1911). There is little evidence that the estuary has increased its trophic status and at present Williams (1979) considers that the calculated average annual productivity for the estuary waters of 123 gC $m^{-2}y^{-1}$ (Bryan 1979) is typical of temperate estuaries.

The effect of an increase in nutrient availability in the estuary could be important. There is already some evidence (Hugman and Mallet, 1976) that the collapse of the peak phytoplankton bloom can contribute significantly to the summer dissolved oxygen sag in the Itchen estuary and possibly in the Test estuary (SWA unpublished data). Increases in the quantity of algal mats appearing on the mudflats can change the underlying fauna (Soulsby et al, 1982) and may lead to changes in the bird population exploiting them (Tubbs, 1977).

The nutrient and metal surveys were designed to produce data which could be used as a baseline against which future surveys could be measured to detect any significant trends. Examination of the means using a least significant range test Student-Neuman-Keuls (Table 2) demonstrates that in general the estuary can be represented

Table 5 Dilution of individual industrial discharges in each of 5 segments of Southampton Water. Dilutions calculated assuming complete mixing and conservative behaviour.

| Discharge | Percent Effluent at Steady State | | | | |
|---|---|---|---|---|---|
| | A | B | C | D | E |
| 1 | 0.04 | 0.05 | 0.03 | 0.03 | 0.01 |
| 2 | 1.01 | 1.11 | 1.17 | 0.71 | 0.65 |
| 3 | 0.01 | 0.01 | 0.01 | <0.01 | <0.01 |
| 4 | 1.11 | 1.23 | 1.33 | 1.19 | 0.58 |
| 5 | 1.99 | 2.19 | 2.37 | 2.45 | 1.08 |
| 6 | 0.03 | 0.03 | 0.04 | 0.04 | 0.02 |
| | Head | | | | Mouth |

by 2 sampling points at the head and the mouth. However, an important consideration for any future monitoring systems is the effort required to determine a realistic change in the means as being significant. Using the means as shown in Table 2, calculation of the number of samples required to detect a 25% or 50% change in the mean as significant ($P < 0.05$) using the method of Montgomery and Hart (1974) (Table 3), suggest that only a sampling regime detecting the latter is feasible. The natural variation of both nutrient and metal concentrations in the system may consequently determine that control is exercised by close examination of the inputs and effects since biological effect may be experienced before the chemical monitoring system is capable of detecting its likelihood.

## DISCUSSION

The range of studies so far undertaken has been at minimal cost and effort. Table 4 shows the average annual number of man days spend on Southampton Water studies to be 65 and total costs at 1981 prices including salaries, boat hire and equipment depreciation to be approximately 5,000 pounds. The information obtained has improved the definition of the overall water quality status of the estuary and perhaps most importantly has identified the variation within the studied parameters and quantified the degree of change which must take place before they can be recognised as significant. The level of survey is capable of providing information for international obligations and central government requirements.

The management of estuarine quality through inputs control will require detailed information on the effect or potential effect of any particular discharge before a decision on changing it in any necessary manner can be taken. It is unlikely that low level studies such as described will provide that information. They will provide a framework on which more detailed study can be based if considered necessary or indicate discharges requiring further investigation.

The present studies have not revealed any positive correlation between specific discharges and effects but do suggest that it is the action and input of the industrial discharges on the western side of the estuary rather than sewage and river inputs at its head that are of concern. The studies have also not differentiated between the present effects of these discharges and the effects of the historical contamination of the sediments.

To provide information of these necessitates a further input of manpower resource since the requirement to detect trends demands the continuation of at least part of the previous programmes. Preliminary studies have commenced on 2 investigations; examination of metal levels in the brown seaweed *Fucus vesiculosus* and the hard shelled clam *Mercenaria mercenaria* and toxicity studies on the industrial effluents. Analysis of the sediments for hydrocarbons will become part of the routine sediment programme.

The metal levels in biota are designed to detect differences in metal burdens and thus metal availability over the whole area of the estuary although they are hampered by the absence of suitable material from the area of concern on the western side. Results so far consist only of 2 surveys of *Fucus* and 1 of *Mercenaria* and are therefore not published here.

Toxicity testing requires specialised facilities with a subsequent increase in manpower and financial investment. Such an approach consequently necessitates careful consideration of the information required. In the case of Southampton Water, the present potential for environmental effect of several large industrial discharges is unknown though substantial investment in effluent treatment has been made in some instances. A useful evaluation of the potential problem is obtained by examining the dilution available to the discharges. Construction of a simple 5-box dilution model using flushing times based on salinity calculated for the estuary by Westwood (1979) and validated using figures for ammonia in winter derived from the present study, demonstrates (Table 5) that the larger effluents could be present at steady state as up to 2.4% of a segment volume. The large volume discharges therefore would have to exhibit comparatively low individual toxicities to be considered 'safe'. The discharges would rapidly come into contact with each other with the possibility of synergistic effects. A comprehensive toxicity survey accounting for the range of biota levels found in the estuary, the inevitable variability of the effluent characteristics and their potential combination in the field would not be feasible within the framework of the present approach. However, it is considered that an examination of the direct toxicities of the larger effluents to one or more test animals would contribute useful information on factors affecting the present water quality status of the estuary. Preliminary investigations along this line have been initiated.

ACKNOWLEDGEMENTS

The Authors wish to thank Mr. H. Speight, Director of Resource Planning of the Southern Water Authority, for permission to write this paper, and Mr. K. Guiver for his advice and comments. Our thanks are also due to Mr. M.J. Beard for chemical analysis and Mr. I.C. Hart for advice on data analysis.

The views expressed are the Authors' own and are not necessarily to be regarded as those of the Authority.

## REFERENCES

Barnes,R.S.K. 1973. The intertidal lamellibranchs of Southampton Water with particular reference to *Cerastoderma edule* and *C. glaucum*. *Proceedings of the Malacological Society, London*, *40*, 413-433.

Barnes,R.S.K., Coughlan, J. and Holmes, N.J. 1973. A preliminary survey of the macroscopic fauna of the Solent, with particular reference to *Crepidula fornicata* and *Ostrea edulis*. *Proceedings of the Malacological Society of London*, *40*, 253-275.

Bryan, *The production and decomposition of organic material in an estuary* - Southampton Water. Ph.D. Thesis, University of Southampton.

Coughlan,J. 1977. Marine Wood Borers in Southampton Water, 1951-75. *Proceedings of Hants Field Club and Archaelogical Society*, *33*, 5-15.

George,J.D. 1964. On some environmental factors affecting the distribution of *Cirriformia tentaculata* (Polychaeta) at Hamble. *Journal of the Marine Biological Association of the United Kingdom*, *44*, 373-388.

Gray,J.S. 1979. The development of a monitoring program for Norway's Coastal Marine Fauna. *Ambio* *8*, 4 176-179.

Hibbert,C.J. 1976. Biomass and production of a bivalve community on an intertidal mudflat. *Journal of Experimental Marine Biology and Ecology*, *25*, 249-2661.

Holmes,N.J. & Coughlan, J. 1975. The ascidian fauna of Southampton Water. *Proceedings of Hants Field Club and Archeological Society*, *30*, 9-15.

Houston,M.C.M., Lowthion, D. and Soulsby, P.G. (in press) The identification and evaluation of benthic macroinvertebrate assemblages in an industrialised estuary - Southampton Water, U.K., using a long term level sampling strategy. Marine Environmental Research - In press.

Hugman, S.J. and Mallet, M.J. 1979. Photosynthesis in the Itchen Estuary, 1974-75. Water Research Centre Technical Report, TR 112, 1979.

Knap, A.H. 1972. A study of the zinc levels in the Solent and Southampton Water with special reference to its uptake by Crepidula Fornicata. M.Sc., University of Southampton.

Knap, A.H. 1979. The fate of non-volatile petroleum hydrocarbons in refinery effluent entering Southampton Water. Ph.D. Thesis, University of Southampton.

Knap, A.H., Williams, P.J. le B. and Tyler, I. 1979. Contributions of volatile petroleum hydrocarbons to the organic carbon budget of an estuary. Nature, 279, 5713, 517-519.

Leatherland, T.M. and Burton, J.D. 1974. The occurrence of some trace metals in coastal organisms with particular reference to the Solent region. Journal of the Marine Biological Association of the United Kingdom, 54, 457-468.

Levell, D. 1979. Southampton Water benthic surveys (1975 and 1976). Field Studies Council, Oil Pollution Research Unit, Pembroke, U.K.

Mackay, D.W. 1975. Techniques for Pollution Control in Estuarial Waters. In: Pollution Criteria for Estuaries, (Helliwell, P.R. and Bossanyi, J. (Eds.)) Pentech Press, London.

Maurer, D., Watling, L., Bottom, D. and Pembroke, A. 1978. Seasonal fluctuation of water quality (nutrients and pigments) in lower Delaware Bay. Hydrobiologia, 60, 3, 203-211,

McLusky, D.S. 1982. The impact of petrochemical effluent on the fauna of an estuarine mudflat. Estuarine Coastal and Shelf Science, 14, 5, 489 - 501.

Mitchell, R. 1974. Aspects of the ecology of the lamellibranch Mercenaria mercenaria in British Waters. Hydrobiological Bulletin 8, 124-138.

Montgomery, H.A.C. and Hart, I.C. 1974. The design of sampling programmes for rivers and effluents. Water Pollution Control, 73, 1, 77-.

N.E.R.C. 1980. The Solent Estuarine System. An Assessment of present knowledge. The Natural Environment Research Council Publications, Series C, No. 22, 1980.

Oyenekan, J.A. 1979. Community structure and production of the benthic macro-infauna of Southampton Water. Ph.D. Thesis, Universty of Southampton.

Phillips, A.J. 1980. Distribution of chemical species. In: The Solent Estuarine System, N.E.R.C. Publications, Series C, No. 22, 44-61.

Raymont, J.E.G. 1972. Some aspects of pollution in Southampton Water. Proceedings of the Royal Society of London, Series B, 180, 451-468.

Romeril,M.G. 1974. Trace metals in sediments and bivalve Mollusca in Southampton Water and the Solent. Revue Internationale d'Oceanographie Mediterranne 33, 31-47.

Romeril,M.G. 1979. The occurrence of copper, iron and zinc in the hard shell clam, *Mercenaria mercenaria* and sediments of Southampton Water. Estuarine and Coastal Marine Science, 9, 423-434.

Royal Commission 1911. Nuisances due to excessive growth of green seaweeds in sewage polluted estuaries with special reference to Belfast Lough. Seventh Report of the Royal Commission on Sewage Disposal. Wyman and Sons, London.

Soulsby,P.G., Lowthion, D. and Houston, M.C.M., 1982. Effects of macroalgal mats on the ecology of intertidal mudflats. Marine Pollution Bulletin, 13, 5, 162-166.

Souza Lima,H. de, and Williams, P.J. le B. 1978. Oxygen consumption by the planktonic population of an estuary - Southampton Water. Estuarine and Coastal Marine Science, 6, 515-523.

Tubbs,C.R. 1977. Wildfowl and waders in Langstone Harbour. British Birds, 70, 177 - 199.

Webber,N.B. 1975. The tidal hydraulics of the Solent and its estuaries. In Pollution Criteria for Estuaries, Helliwell, P.R. and Bossanyi, J. (Eds). Pentech Press, London.

Webber,N.B. 1980. Hydrography and Water Circulation in the Solent. In: The Solent Estuarine System. N.E.R.C., Huntingdon Publications, Series C, No. 22, 25-36.

Westwood,I.J. 1979. Mixing and dispersion in Southampton Water. Ph.D. Thesis, University of Southampton.

Williams,P.J. le B. 1979. Primary productivity and heterotrophic activity in estuaries. River Inputs into Ocean Systems, Rome, 1979.

Wright,S.L. 1980. The pollution load entering Southampton Water and the Solent. In: The Solent Estuarine System. N.E.R.C. Publications, Series C, No. 22, 62 - 64.

# PREDICTIVE MODELLING IN ESTUARY QUALITY MANAGEMENT

P.C. Nicholas

Welsh Water Authority
Tidal Water Authority
Tidal Waters Unit
Tremains House
Coychurch Rd.
Bridgend CF31 2AR
Wales

## INTRODUCTION

During the past 25 years problems associated with pollution of estuaries by industrial and domestic effluent have caused increasing concern. This has been reflected in the increase in the pollution control legislation in the United Kingdom.

The water quality management of estuarine and certain coastal waters of England and Wales is vested in the Regional Water Authorities as set up and defined by the Water Act, 1973. This Act incorporated certain of the recommendations of the Third Report of the Royal Commission on Environmental Pollution (1972).

The Water Authorities operate within the powers of relevant statutes. In particular, effluent discharges to "controlled" tidal waters are subject to controls outlined in the Rivers (Prevention of Pollution) Acts, 1951 - 1961 and in the Clean Rivers (Estuaries and Tidal Waters) Act, 1960. The latter designates major estuaries and certain coastal waters as "controlled" and gives powers to impose discharge consent conditions to any new or altered outlets.

These powers will be further extended by the implementation of the Control of Pollution Act 1974. The Act will extend "controlled" waters to a 3 mile coastal limit and will provide powers to control the quality of existing discharges. Although the 1974 Act will repeal previous statutes and set a more formal and publicly accessible procedure for consents, the basic powers remain generally unchanged.

Whilst the Pollution Control Acts provide the basic legislative powers, management is also governed by the Public Health (Drainage of Trade Premises Act), 1936, the Water Act, 1973, the Wildlife and Countryside Act, 1981, the Dumping at Sea Act, 1974 and the Salmon and Freshwater Fisheries Act, 1975.

These statutes provide the machinery for pollution control, and in some cases provide broad objectives, but neither environmental quality standards nor emission standards for effluents are defined.

EEC Directives have the status of domestic legislation, and offer alternatives between fixed emission standards and environmental quality standards (EQS). The U.K. favours the EQS approach. Domestic legislation is used as the machinery for the implementation of these Directives. EEC Directives relating to Welsh estuaries which have been implemented include:

1. Council Directive (76/160/EEC) concerning the quality of Bathing Water (1976);
2. Council Directive (76/464/EEC) on Pollution caused by certain Dangerous Substances Discharged to the Aquatic Environment of the Community (1976);
3. Council Directive (82/176/EEC) on Mercury Discharges by the Chlor-alkali Electrolysis Industry (1982);
4. Council Directive (79/923/EEC) on the Quality Required of Shellfish Waters (1979);
5. Council Directive (83/513/EEC) on Limit Values and Quality Objectives for Cadmium Discharges (1983).

In addition, the Council Directive on the Discharge of Mercury by Sectors other than the Chlor-alkali Industry awaits implementation.

The Regional Water Authority, in discharging its various statutory responsibilities and being itself a major discharger, must recognise the cost effectiveness of effluent disposal to tidal waters. Equally, the same Authority must discharge its various statutory responsibilities for protection of the aquatic environment. Between these opposing objectives lies a central position which is defined in more detail by Authority statements of policy. Some latitude between the basic statutes and their implementation exists and, therefore, statements of policy will differ between Authorities.

The policies of the Welsh Water Authority are based on the recommendations of the Working Party on Sewage Disposal (1971) and the Royal Commission on Environmental Pollution (1973). These policies have been stated by Howells et al (1978) and provide for the cost effective disposal of wastes to estuaries, whilst retaining a commitment to maintain standards of amenity and so protect aquatic life. Thus, the Welsh Water Authority (WWA) seeks to establish a balance between cost effective exploitation of inshore waters whilst ensuring the protection of the nearshore marine environment.

Although the mechanisms necessary for pollution control are contained in the United Kingdom legislation, the relevant statutes do not specify quality standards for either effluents or receiving waters. It is not United Kingdom government policy to specify uniform quality standards. Standards are set in conformity with the uses of the receiving waters, either on a national or local basis. In fact, the only national standards are the receiving water quality standards specified by EEC Directives, or set by Government in compliance with these Directives. Effluent quality control standards are then derived by the competent Regional Authorities. When dealing with pollutants or water uses not covered by EEC Directives, the regional authorities will set receiving water quality standards where-ever necessary, with their consequential effluent control standards.

This approach ensures not only a proper standard of protection for the receiving waters but also the most economical use of these waters for effluent disposal. This contrasts with the use of emission standards, fixed on the basis of available means of treatment or in the nature of the pollutant concerned.

The use of the environmental quality standard approach has one major disadvantage. It requires a knowledge of existing quality, the effects of pollutants on the biota and the complex interrelationships which control these. Consequently, either lengthy monitoring or short term intensive studies of the receiving water and the relevant pollutants must be carried out. In the complex regime encountered in tidal waters, experience indicates that the latter course is most effective, both in terms of the quality of information obtained and of cost.

In order that best use of this information can be made, linking discharges with effects and assessing the implications of change, there is no question that the complex interactions which can occur are most adequately accounted for by the technique of predictive modelling. The Welsh Water Authority are committed to this approach and have developed predictive models for the Usk, Dee, Tywi, Conwy, Milford Haven, Neath and Severn Estuaries.

This paper describes the application of mathematical models to estuary quality management problems within Wales.

## THE BASIS OF PREDICTIVE MODELLING

Because of the demands on modelling resources and the number and diversity of impending problems, each model has been based upon the simplest representation of the primary physical and biochemical processes which provide realistic prediction. This is a necessary requirement since, for example, the statutory response deadline to a consent applicatin is 3 months. In other instances, where schemes are considered on an engineering or operational basis (e.g. surcharged sewers or local flooding), it is imperative that environmental impact appraisals are complete at an early stage if they are to be given due consideration in the decision making process.

In short, in the working environment of the Water Industry, the requirement is often driven by the need to use the best available means of assessment within a given reporting deadline. If the deadline permits the production or application of a predictive model, it is vital that the basic formulation of the model is checked for correctness, and that the consequences of simplifying assumptions are tested.

The procedure of validation by which the predictive accuracy and sensitivity of a model is assessed against observation is a time consuming and expensive problem.

Whilst many models appear in the published literature, it is usual to find that only the hydrodynamic aspects of these models are well validated against records of tidal elevation, tidal velocity or both. Conversely, dispersion and water quality models are often only poorly, or incompletely validated. There are perhaps several reasons for this:

1. The hydrodynamic model is often developed as a reseach tool in departments of civil or chemical engineering, physical oceanography or research institutes. In many cases, the academic objectives such as a new numerical solution, or the inclusion of an additional term such as wind shear will be satisfied by the proven hydrodynamic model;

2. Engineering consultants add to the modelling effort, but again commercial contracts are often satisfied by a proven hydrodynamic model, for instance in the design of sea defences or barrages;

3. Few organisations have the blend of expertise, resources and research or commercial interest to implement and validate a water quality model. As a direct consequence, it is usual to find a water quality or dispersion model checked against limited observation and based upon poorly described kinetic-rate equation;

4. Whilst the water quality model is of undoubted value in the investigation of processes involved in marine water quality, research has resulted in little development or new understanding of kinetic processes over the past decade, and these accepted values are often based on a very few observations. There is,

therefore, a tendency to adjust the ratio of these processes as a variable in the calibration of the model, leading to confusion between this procedure and true validation.

Within this climate, the modeller has to recommend effluent consent standards. Whilst the techniques and existing knowledge are far from perfect, it must be acknowledged that the predictive model is a considerable refinement upon simplistic calculation, such as pollution budgeting.

What determines a validated model? Certainly no written rules or codes of practice exist, and the degree of rigour exercised is inevitably determined by the modeller. This will be limited by time, data base, costs, knowledge, computing capacity or, indeed, the particular characteristics of the subject estuary. The process of validation can be broken down into three stages:

## Analytical Solution

The application of the predictive model to a simple hydraulic problem for which there is a known analytical solution is a valuable technique for proving the correctness of the computation procedure. The model is, however, being used with all its simplifying assumptions and mathematics to solve a hypothetical set of circumstances. This procedure cannot be regarded as confirmation that the same simplifying assumptions apply to the estuary on which it is intended to apply the model.

The procedure for validation may vary in detail considerable between estuaries, with the model used, and the parameter predicted. There are, however, important underlying rules which are generally applied.

## Calibration

The most important aspect of validation is to separate the data used by the model, such as volumes or river flows, from the observation against which the check is to be carried out. If the model is capable of this independent prediction, then it is possible to establish basic confidence in its predictive accuracy. If the model is not capable of independent prediction, the method must be regarded as stochastic and, strictly speaking, not predictive.

If we assume that the prediction and observation are in close agreement, can the model be considered validated? If we consider the combined potential errors of all data used by the model, then it is clear that accurate prediction can only be regarded as fortuitous. This is more easily appreciated by the usual case where model prediction and observation shown some disagreement. It is the common practice here to adjust data or rate constants, within justifiable limits, until a "best fit" is produced.

Whether the first prediction is correct, or whether minor adjustment is necessary, the procedure must be regarded as a calibration of the model.

## Validation

The model is designed to predict change and, in particular, to predict a condition which cannot otherwise be monitored. It is, therefore, important to check the accuracy of prediction against the largest range of conditions possible. These are of typically, repeat predictions for neap and spring tides, a range of river run off conditions and, where possible, for a major change in effluent loadings. The combinations of tests under a variety of conditions improves confidence in validation. This level of rigour is not always justifiable but an example is shown for the Usk Estuary in Figure 1. The ease with which validation can be achieved varies with the quality conditions which prevail in the receiving waters:

a. Polluted Estuaries. The data presented in Figure 1 indicate a significant level of pollution under low flow conditions. The trend of an increasing oxygen sag under reducing river flow allows sensitive calibration and validation of the model.

b. Grossly Polluted Estuaries. Often, the change between the polluted and grossly polluted case involves a change in kinetic processes. The step change from aerobic to anaerobic conditions is known. The exact change in rate constants and the varying role of sediments is particularly poorly understood. This poses particular problems for prediction of changes in water quality.

c. Clean Estuaries. Surprisingly, the unpolluted estuary provides a equally difficult problem of validation. The problem is to isolate a significant trend in water quality which is not merely a trivial interpolation between river and sea boundary conditions. Prediction of this condition can, of course, be achieved but this provides little confidence in the accuracy of the model.

   In such circumstances, it may be necessary to introduce an artificial change which can be monitored. The introduction of a fluorescent dye, radioactive source, bacteria or chemical tracer are all viable techniques. The method of release and monitoring can be specifically tailored to the predictive capability of the model. An example is shown as Figure 2.

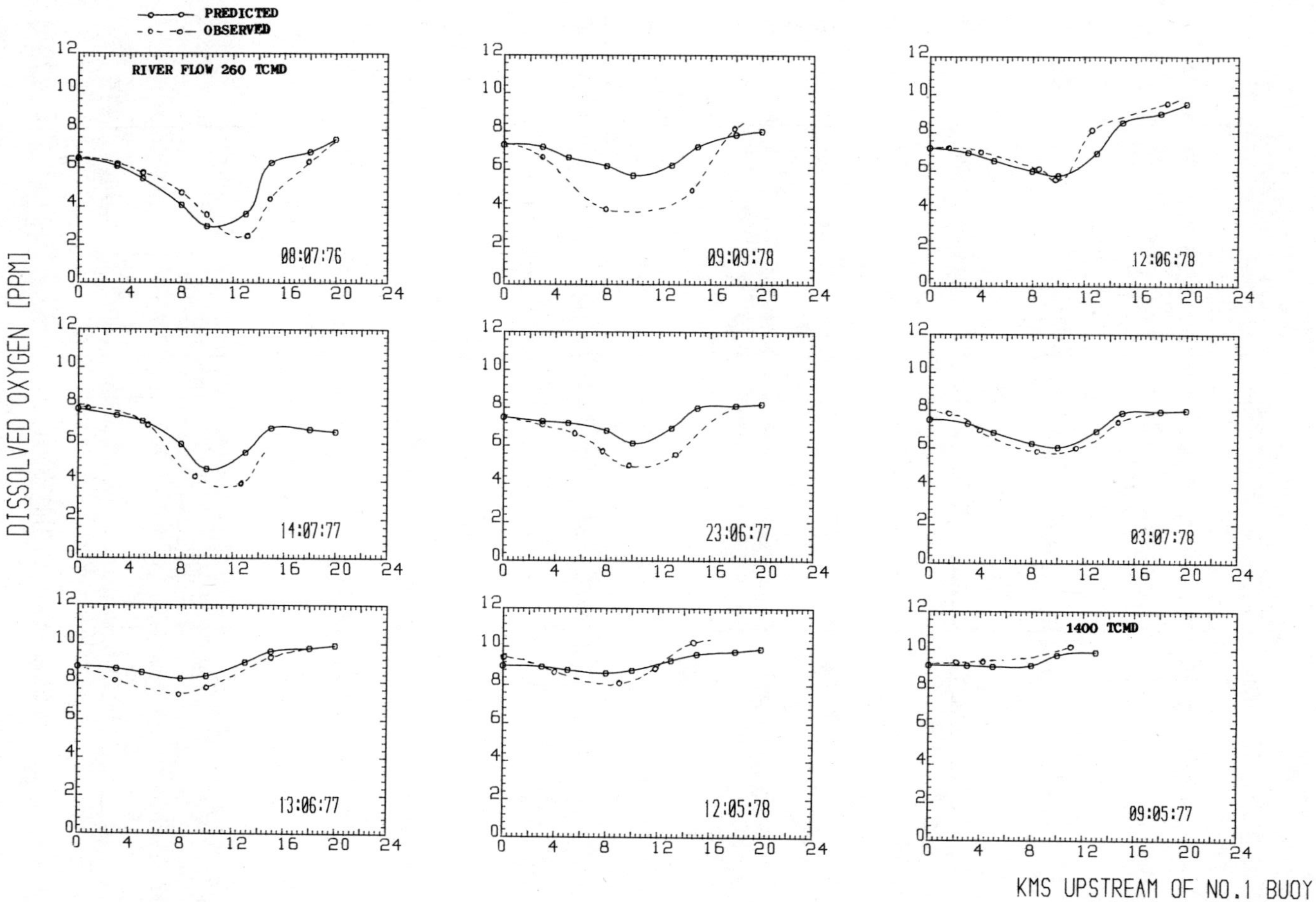


Fig. 1.1 Validation of spring tide D.O. model over a large range of river flows, showing tidally averaged depth mean profiles presented with moving segments in their mid ebb positions.

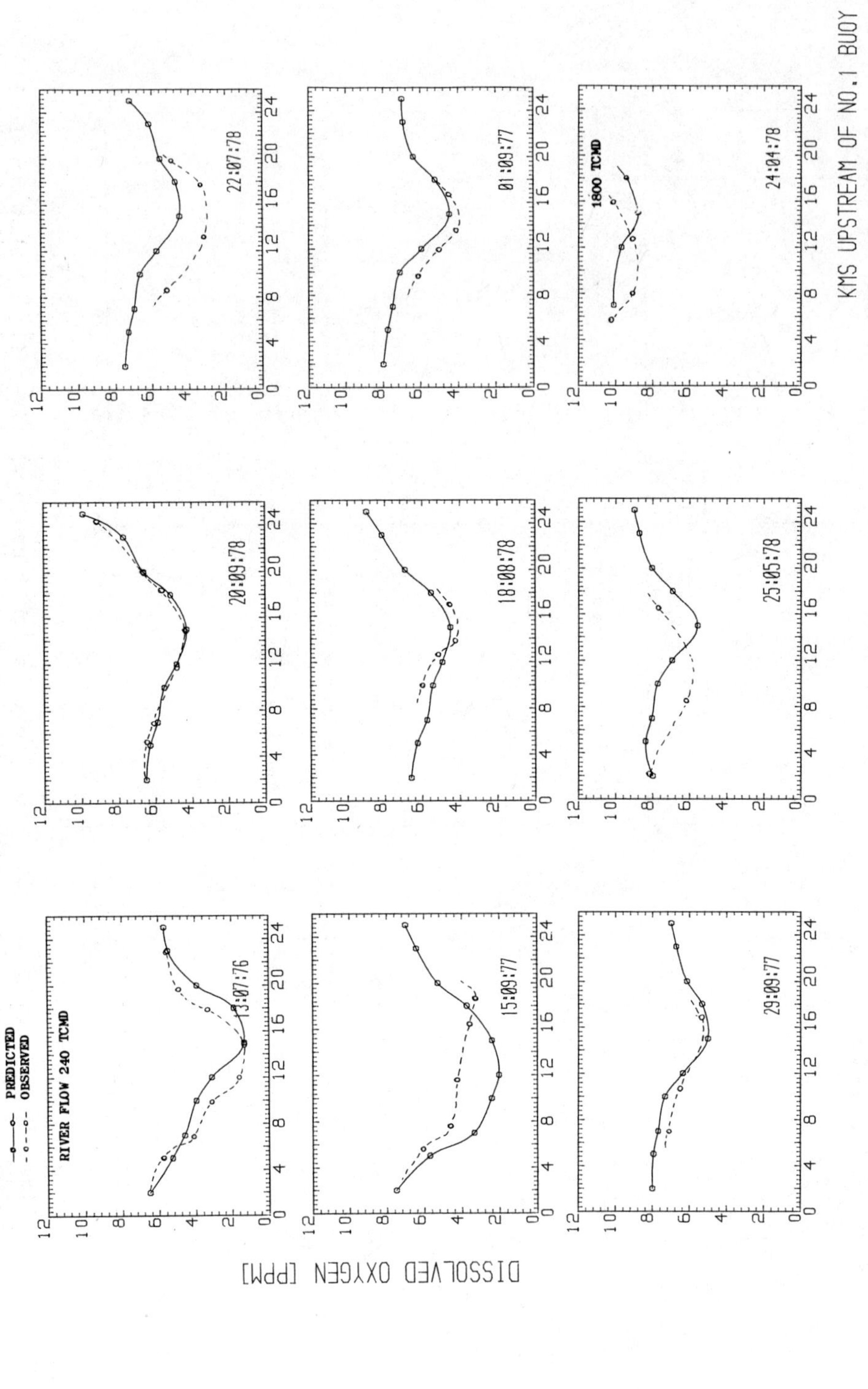


Fig. 1.2 Validation of neap tide D.O. model. Notation as Fig. 1.2.

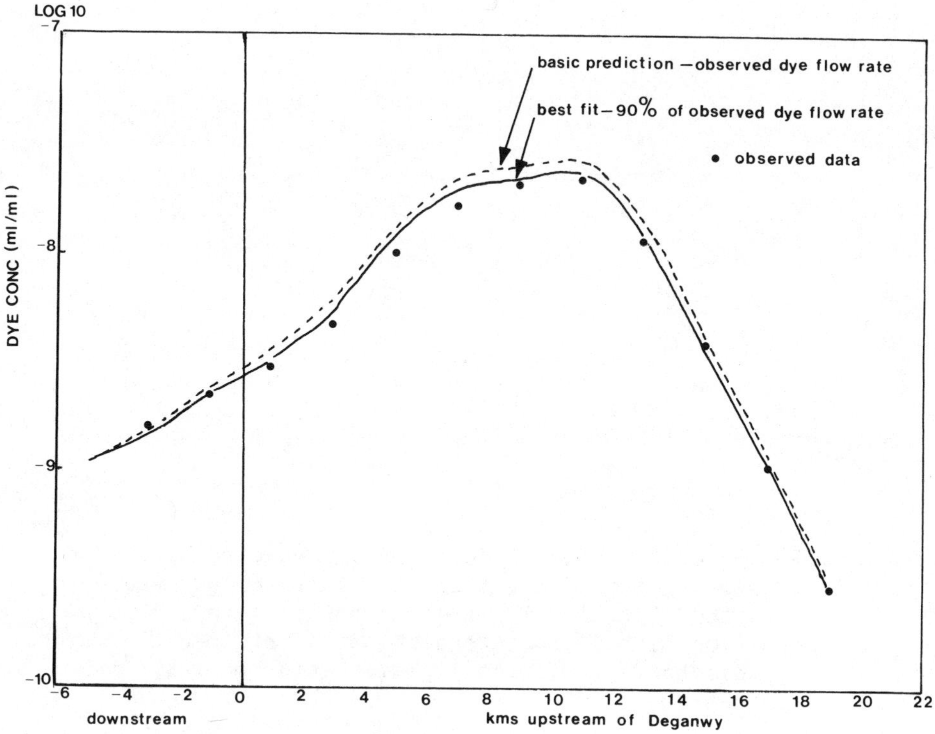


Fig. 2 Continuous dye release predicted vs observed data -spring tide

APPLICATIONS OF PREDICTIVE MODELLING

The objective of developing and using a predictive mathematical model is often the assessment of alternative disposal options for a major drainage scheme. The model can be applied directly, and a typical output is shown as Figure 3. This illustrates an existing condition, an equivalent prediction and the net benefit after implementing alternative schemes. Equally, the same model may be used to assess the deterioration by allowing new discharges. Figure 4 illustrates the typical case where spare capacity exists between the existing condition and the quality standard and where part of this capacity could be allocated to absorb the new pollutant load.

In practice, the interpretation of such results tends to vary with time, subject to prevailing circumstances. For instance, the 1976 drought created a stringent attitude to consent setting for some 2 years. This view has since been superseded by a more relaxed approach in recognition of economic circumstances and industrial problems, related to low river flow conditions which occur more frequently than the extreme cases observed in 1976.

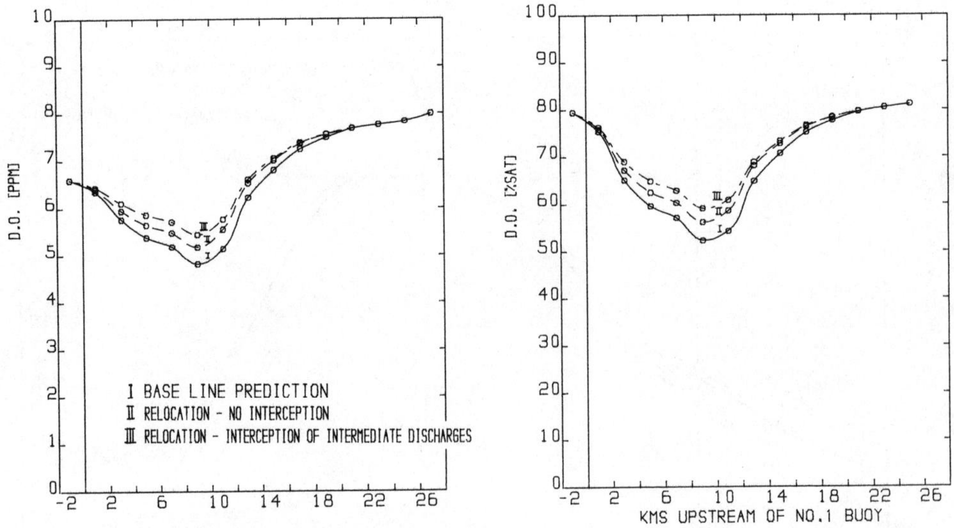


Fig. 3 Relocating Ponthir effluent: predicted effects on spring tide D.O. Solid line - baseline prediction, Ponthir STW on-line; dashed line - relocation with 1) no interception of intermediate discharges; 2) part implementation of West Bank scheme.

This one-off, or *ad hoc* application, of the predictive model is wasteful of the potential resource. In particular, if decisions are made with primary regard to external pressures, the result is inevitably a sequence of inequitable consents. This, in turn, leads to poor relations between the Water Authority and Industry.

The objective is, quite clear - namely, to set up a framework within which consents can be assessed equally over a sustained period of time and within which the need to implement remedial schemes can be critically reviewed. Alternative methods for setting up such a framework all fall into the realm of planning and policy.

## Fixed Emission Standards

The use of predictive modelling within such a policy is limited as fixed emission consent would limit the quality of effluent discharged to a specified limit. International legislation adopts this method for certain persistent or toxic substances and severely limits their discharge. Within this context, the fixed standard applied to industry on a large scale, nationally or internationally, has merit.

It is possible to postulate the application of such standards to degradable and non-persistent discharges. The advantage is the ease and consistency of application.

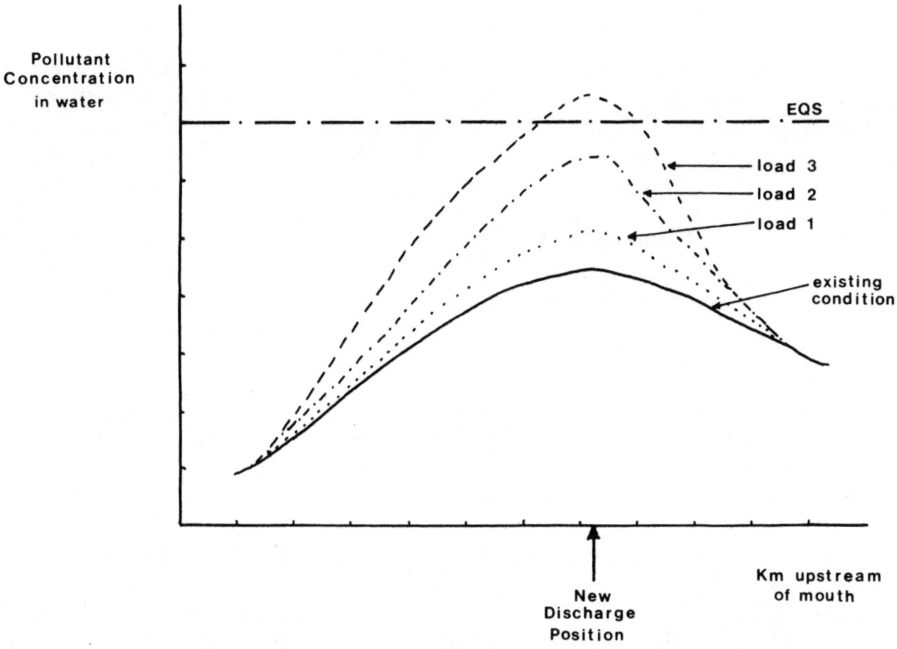


Fig. 4 Predicted effects on water quality of a new discharge for varying loads of pollutant

Applied simply, the method would fail to recognize the variation in discharge capacity between the mouth and head of an estuary. Applied to several estuaries, orders of magnitude difference would exist in capacity to absorb and degrade effluent. This is wasteful of the marine resource as an economic means of effluent disposal and would not comply with stated policy.

Emission standards could be tuned to the individual receiving water. This could be achieved by establishing capacity between existing conditions and a quality standard for individual zones of an estuary. Taking one such zone, any effluent discharge would receive a identical quality consent. Before determining that consent, the potential rate of industrial development (or closure) would have to be estimated for some foreseeable period and the capacity or part capacity reserved accordingly.

The procedure may be criticised since, if industrial development does not materialise or if closures continue, those consents granted will be over-restrictive. Further, the method does not recognise the relative difficulty and cost of treating certain effluents. Equally, if growth, albeit unlikely, exceeds expectations, the method fails to offer environmental protection. In essence, this procedure fails to achieve stated policy.

## Assessment of Individual Discharges

There remains considerable cost benefit in considering individual discharges based upon the point of discharge, tidal conditions and ease of treatment. Intangible factors of a political nature, such as jobs created and benefit to the local economy, may also have significant influence. Whilst these should be considered in appropriate cases, these secondary factors should be constrained. This can be achieved by establishing a formal consenting method to operate within, analagous to a code of practice. The following methods are proposed to further this discussion.

a. Fixed Assessment of Loading Consent

In this alternative, capacity for new discharge is assessed at a base date. A proportion of this capacity is then allocated to each new discharge, the consent being derived on an acceptable load basis rather than an emission standard basis.

The case for flexible allocation of capacity dependent on, for example discharge site, remains. This approach would normally result in identical consents for adjacent factories with similar effluents, irrespective of application date. It does, however, permit a variable allocation of capacity based upon the best practical means of treatment.

This method offers a consistent approach to the consenting officer. It does not, however, make allowance for periods of significant industrial closure since the available disposal capacity is not recalculated. Neither does it protect the environment during an underestimated phase of industrial growth without the imposition of zero emission standards.

b. Updated Assessment of Loading Consents

The above limitation can be overcome by assessing capacity at the time of the consent allocation. This requires the existing position to be continually updated depending upon changes to, for example, effluent quality, policy and industrial closures. Consents set will not, therefore, be consistent, for otherwise identical discharges, over a period of time, but will depend upon prevailing circumstances.

The method requires continually updating after each change in effluent standards and for each new scheme or discharge. Perhaps the major argument against continually updating is the work and delay time involved in producing results. This can be overcome by producing a generalised package which predicts a variety of single changes to the modelled estuary. Summation of predictions for multiple changes holds good for short periods although clearly, for major multiple changes, the full model must be applied.

## LONG TERM PLANNING IN RELATION TO ENVIRONMENTAL QUALITY STANDARDS

The strategy developed so far concentrates upon the setting of a sequence of consents. This amounts to 'controlled deterioration' of the receiving water quality which can be allowed down to a specified environmental quality standard (EQS). Further deterioration must be viewed as unacceptable and there can be considerable practical difficulty in determining consents. The method advocated, namely, updated assessment of loading consents, would result in progressively more stringent consents being set as the EQS is approached. This will eventually prohibit new discharges on economic grounds.

A logical extension of the strategy is to incorporate an Authority's own improvement schemes, often the major source of pollution, into the projected plan. In essence, deterioration is allowed to occur until an EQS is reached, at which time, immediate implementation of an improvement scheme is necessary. With some basic estimates of development rate, it is possible to plan the timescale for Authority capital expenditure. The case discussed so far is a clean estuary with identified capacity for new discharge. The progressive changes are illustrated schematically as Figure 5(i) and the objective is quite clearly to maintain standards above the defined EQS.

In a more polluted estuary, the same basic proposals apply. Existing water quality will be below the desired standard (Figure 5(ii). Here, the desired quality can be interpreted as an environmental quality objective (EQO). The objective can, in the usual case, only be achieved by the implementation of capital schemes. The objective must, however, compete for priority expenditure against schemes concerned with water supply, and public health and it is not unusual to find schemes for the improvement of tidal waters afforded a low priority.

Within this climate, where improvement, although perhaps agreed and quantified, is some years away, the Authority must continue to determine consents. A policy of no deterioration is often difficult to justify and the acceptance of minor deterioration is almost inevitable.

By establishing a minimum acceptable standard for an interim period, the capacity for minor deterioration can be quantified. It is important to recognise in setting such a standard, an EQS, that its purpose is to protect existing use. A more stringent quality objective, EQO, can be retained, this defines the levels of improvement necessary for potential new uses of the receiving waters. Simultaneous use of the quality objective and quality standard may prove

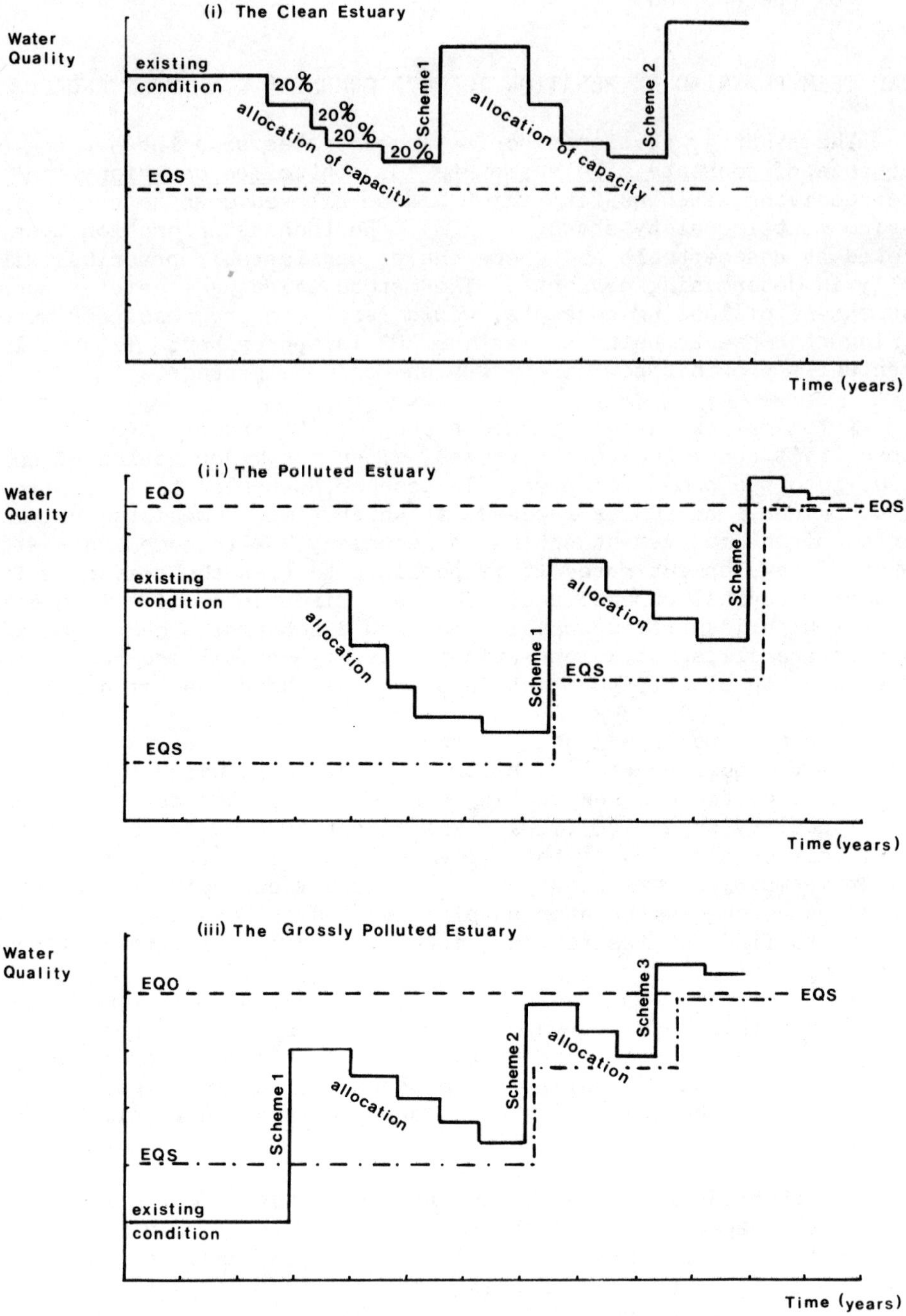


Fig. 5 Schematic illustration of management in estuaries

confusing to some readers and it is important to appreciate the distinctly different definition being used within the context of this paper.

Consents based upon the capacity between the existing condition and EQS would, therefore, proceed as before. A more stringent attitude towards each consent should be adopted and a consent, if allocated, should not ensure that a best practical means of treatment is carried out as a minimum level of compliance. Achievement of the long term objective would be dependent upon the prevailing economic climate.

Only in cases of severe pollution, Figure 5(iii), where water quality fails to meet the minimum acceptable standard, can a policy of no deterioration be justified. Improvement schemes should be considered a matter of priority and improvement to a level giving spare disposal capacity in excess of the minimum EQS compliance is required. The more stringent quality objective still remains but must be considered a matter of lower priority. The position would revert to the case of a polluted estuary considered previously.

## CONCLUSIONS

Mathematical models are already in use on many significant estuaries within Britain. Their use provides management and the research scientist with a vastly improved quality of information.

It is felt that a major proportion of model application work is carried out on an *ad hoc* basis. Extension of this form of work to a more formalised strategic plan is seen as an important development. This allows a consistent framework for the setting of industrial and Authority consents. This framework can be combined with programmes for the implementation of the Authority's capital schemes. This, in turn, allows for the early identification and allocation of the capital budgets.

This paper can only go some way to recommending guidelines specifying the quality of work which constitutes a validated model but, more particularly, the way in which the predictive aid can be applied to identify and to allocate effluent disposal capacity within the framework of a long term plan.

## ACKNOWLEDGEMENTS

The author wishes to thank E.J. Bent and Mr. W. Halcrow who critically examined the manuscript. Thanks are due also to the Welsh Water Authority for permission to publish this paper.

The views in this paper are those of the author and do not necessarily represent those of the Welsh Water Authority.

REFERENCES

EEC, 1976a. Council Directive (76/160/EEC) concerning the Quality of Bathing Water 1976. Official Journal of the European Community. L31/1.

EEC, 1976b, Council Directive (76/464/EEC) on Pollutin caused by Certain Dangerous Substances Discharged to the Aquatic Environment of the Community, 1976. Official Journal of the European Community. L129/23.

EEC, 1979, Council Directive (79/923/EEC) on the Quality Required of Shellfish Waters, 1979. Official Journal of the European Community L281/47.

EEC 1982, Council Directive (82/176/EEC) on Mercury Discharges by the Chlor-Alkali Electrolysis Industry, 1982. Official Journal of the European Community L81/29.

EEC, 1983, Council Directive (83/513/EEC) on Limit Values and Quality Objectives for Cadmium Discharges, 1983. Official Journal of the European Community, L291/1.

Howells, W.R., Owens M. and Stoner, J.H., 1978. Water Quality Aspects of Welsh Estuarine and Coastal Waters. Journal of the Institution of Water Engineers and Scientists 32 365.

Royal Commission on Environmental Pollution, 1972, Third Report, H.M.S.O., London.

LIST OF CONTRIBUTORS

Barnett, B.E. AWA Lincolnshire River Division, 'The Poplars', Bridge Street, Brigg, South Humberside DN20 8NQ, U.K.

Bettencourt, A., Ecole Normale Superiore, Lab de Geologie, 46 Rue d'Ulm, 75230 Paris Cedex 05, France.

Blackith, Prof. R. Zoology Dept. Trinity College, Dublin 2. Ireland.

Chambers, R.G., NWW Rivers Division, P.O. Box 12, New Town House, Buttermarket Street,, Warrington WA1 2QG, U.K.

Clarke, F., Agricultural Institute, Oak Park, Carlow, Ireland.

Cranshaw, D.H., NWW Rivers Division, P.O. Box 12, New Town House, Buttermarket St., Warrington WA1 2QG, U.K.

Crinion, S., Hill Crest, Amara Estate, Ballina, Co. Mayo, Ireland.

Curran, J.C., Clyde RPB, Rivers House, Murray Rd., East Kilbride, Glasgow G75 OLA, Scotland.

Dowley, A., Engineering Dept. U.C.D., Upper Merrion St., Dublin 2. Ireland.

Ducrotoy, J.P., Station D'Etudes en Baie de Somme, Rue St. Pierre, 80230 St. Valery-sur-Somme, France.

Elliott, M., Forth RPB Lab, Port Edgar Marina, South Queensferry, Edinburgh EH30 9SQ, Scotland.

Essink, K., Ryksinst. voor Zuivering van Afralivater, Lab. Groningen, Hereweg 99a, 972 AA Groningen, Holland.

Evans, M., Welsh Water Authority, Tremains House, Coychurch Rd., Bridgend, Mid;Glamorgan CF31 2AR, Wales.

Halcrow, W., Welsh Water Authority, Tremains House, Coychurch Rd., Bridgend, Mid-Glamorgan CF31 2AR, Wales.

Holmes, P.J., Welsh Water Authority, Tremains House, Coychurch Rd., Bridgend, Mid-Glamorga CF31 2AR, Wales.

Howie, Prof. D.I.D., Zoology Dept. Trinity College, Dublin 2, Ireland.

Hubert, P., Centre d'Information Geologique, Ecole des Mines de Paris, F 77305 Fontainebleau, France.

Jeffrey, D.W., Environmental Sciences Unit, Trinity College, Dublin 2. Ireland.

Khoester, M., Postbus 43g, 4330 AK Middelburg, The Netherlands.

Kilfeather, P.J., Southern Region Fisheries Board, 12 Gladstone St., Clonmel, Ireland.

Langston, W.J., The Laboratory, Citadel Hill, Plymouth PL1 2PB, U.K.
McKay, D.W., Clyde RPB, RIvers House, Murray Rd., East Kilbride, Glasgow G75 OLA, Scotland.
McLusky, D.S., Dept. Biological Science, Sterling University, Sterling, Scotland.
Mansfield, M., An Foras Forbatha, St. Martins House, Waterloo Road, Dublin 4. Ireland.
Maskell, J.A., Hydraulics Research Station, Wallingford, Oxon OX10 8BA, U.K.
Mollowney, M., Water Research Centre, Elder Way, Stevenage, Herts SG1 1TH, U.K.
Moore, D., Dept of the Environment, Customs House, Dublin 1, Ireland.
Nicholas, P., Tremains House, Coychurch Rd., Bridgend, Mid-Glamorgan CF31 2AR, Wales.
Nienhuis, P.H., Vierstraat 28, 4401 EA Yerseke, The Netherlands.
O'Donnell, A.R., Water Research Centre, Elder Way, Stevenage, Herts. SA1 1TH, U.K.
Odd, N.V., Hydraulics Research Station, Wallingford, Oxon OX10 8BA, U.K.
O'Kane, P., Engineering Department, U.C.D., Upper Merrion St., Dublin 2, Ireland.
Osbaldeston, J., North West Water Authority, 54 Brandreth Drive, Parbold, Lancs WN8 7HD, U.K.
O'Sullivan, G., 3 Langford Terrace, Monkstown, Blackrock, Co. Dublin, Ireland.
O'Sullivan, M., Fisheries Research Centre, Abbottstown, Castleknock, Co. Dublin, Ireland.
Parker, J., Dept. Agriculture Fisheries Lab, Castleroe Rd., Coleraine BT51 3RL, Northern Ireland.
Parker, M.M., Fisheries Research Centre, Abbottstown, Castleknock, Co. Dublin. Ireland.
Portmann, J., MAFF, Fisheries Lab., Burnham-on-Crouch, Essex, U.K.
Poulin, M., Centre d'Information Geologique, Ecole des Mines de Paris, F77305 Fontainebleau, France.
Radford, P., IMER, Prospect Place, The Hoe, Plymouth PL1 3DH, U.K.
Read, P., Dept. Biological Science, Napier College, Colinton Rd., Edinburgh EH10 5DT, Scotland.
Riddle, A.M., ICI Brisham Lab, Freshwater Quarry, Brisham, Devon, U.K.
Riordan, D., P.H. McCarthy, Son & Partners, Rosemount Hall, Dundrum Rd., Dublin 14, Ireland.
Rogers, J.B., Southern Region Fisheries Board, 12 Gladston Dr., Clonmel, Ireland.
Rogers, K., 5 Renmore Crescent, Galway, Ireland.
Sayers, D.R., Angglican Water Authority, Ambury Rd., Huntingdon, Cambs. PE18 6N2, U.K.
Soolsby, P.G., Southern Water Authority, Resource Labs, Otherbourne, Hampshire, U.K.
Stapleton, L., An Foras Forbatha, St. Martins House, Waterloo Rd., Dublin 4. Ireland.

Whitelaw, K., Water Research Centre, Henley Rd., Medmenham P.O. Box 16, Marlow, Bucks. SL7 2HD, U.K.
Widdows, J., IMER, Prospect Place, The Hoe, Plymouth PL1 3DH, U.K.
Wilkinson, M., Dept. Brewing & Biological Sciences, Herriot-Watt University, Chambers St., Edinburgh EH1 1HX, Scotland.
Wilson, J.A., Industrial Science Division Dept. Commerce, 17 Antrim Rd., Lisburn, Co. Antrim 3T 28 3AL, Northern Ireland. Ireland.

INDEX